DATE DUE

Digestive physiology and nutrition of marsupials

MONOGRAPHS ON MARSUPIAL BIOLOGY

Digestive physiology and nutrition of marsupials

IAN D. HUME

*Associate Professor, Department of Biochemistry and Nutrition
University of New England, Armidale, Australia*

CAMBRIDGE UNIVERSITY PRESS
Cambridge
London   New York   New Rochelle
Melbourne   Sydney

Published by the Press Syndicate of the University of Cambridge
The Pitt Building, Trumpington Street, Cambridge CB2 1RP
32 East 57th Street, New York, NY 10022, USA
296 Beaconsfield Parade, Middle Park, Melbourne 3206, Australia

© Cambridge University Press 1982

First published 1982

Printed in Great Britain at the
University Press, Cambridge

Library of Congress catalogue card number 81-17032

British Library Cataloguing in Publication Data
Hume, Ian D.
Digestive physiology and nutrition of marsupials.
– (Monographs on marsupial biology)
1. Marsupiala–Physiology
I. Title II. Series
599.2′041   QL737.M3
ISBN 0 521 23892 7

Contents

	Preface	*page* vii
1	**Marsupial metabolism and nutrient requirements**	1
	Basal metabolism and thermoregulation	1
	Variations in basal metabolic rate	5
	The consequences of a low metabolic rate	12
	Conclusion	26
2	**Carnivorous marsupials**	27
	Diet studies	29
	Digestive tract morphology	35
	Digestive function	39
	Specific adaptations in small dasyurid species	41
	Summary and conclusions	49
3	**Bandicoots and other marsupial omnivores**	50
	Bandicoots and bilbies	50
	The Opossums	56
	Omnivorous arboreal marsupials	59
	Conclusions	67
4	**Herbivorous marsupials – the non-macropodids**	69
	Hindgut fermentation	69
	Wombats	70
	The arboreal folivores	75
	The Brushtail Possums	76
	The Koala	83
	The Greater Glider and the Ringtail Possums	97
	Eucalyptus foliage as a food resource	103
	Summary and conclusions	109
5	**Herbivorous marsupials – digestion and metabolism in kangaroos and wallabies**	111
	Digestion in foregut fermenters	111

	The macropodine digestive tract	112
	Regurgitation	117
	Salivary glands	119
	Passage of digesta through the gut	122
	Form and function of the macropodine stomach	127
	Fermentation and microbiology	140
	VFA and carbohydrate metabolism	151
	Lipid metabolism	155
	Nitrogen metabolism and urea recycling	155
	Conclusion	158
6	**Diet and nutrition of kangaroos and wallabies**	159
	Dentition and diet (grazer versus browser)	159
	Nutrition and ecology	168
	Summary and conclusions	199
7	**Herbivorous marsupials – the rat-kangaroos**	200
	Food habits	200
	The digestive tract	204
	Possible evolution of the macropodid digestive system	208
	Conclusions	211
8	**Mineral and vitamin nutrition of marsupials**	212
	Copper, molybdenum and sulphate	213
	Cobalt and vitamin B_{12}	215
	Selenium and vitamin E	215
	Sodium and potassium	217
	Ascorbic acid biosynthesis in marsupials and monotremes and the phylogeny of marsupials	221
	Summary	226
	Appendix	227
	References	231
	Index	253

Preface

Two developments have prompted the writing of this book. The first of these is the rapid growth of interest in and knowledge of general marsupial biology. Although growth in the areas of digestive physiology and nutrition began later than in other areas such as reproductive physiology, there is sufficient information now available from recent and continuing studies in several centres for a monograph on marsupial digestion and nutrition to be written. The central role of food availability in population density and stability has long been recognised. It is hoped that this book will provide the zoologist with an appreciation of the diversity of food resources exploited by marsupials in various habitats, and of specialisations that have evolved to enable these food resources to be exploited.

The second development of note is the increasing awareness among physiologists and nutritionists of the value of comparative studies of different digestive systems. For instance, a prodigious amount of research on the digestive system of domestic ruminants has produced steady progress over the last 35 years in our understanding of that system. But research is never finished, and we still have much to learn about ruminant digestion and metabolism. In recent years has come the idea that in order to understand the ruminant system better, it may be worthwhile to study other foregut fermentation systems as well. The study of the same problem in several different animal species, perhaps only very distantly related but faced with similar environmental challenges, can be expected to lead to alternative ways of looking at the problem in the domestic species. The utilisation of fibrous plant material by macropodid marsupials and by eutherian ruminants is one example of such a problem. It is hoped that this book will provide the nutritionist with an appreciation of the value of comparative studies in digestion and nutrition.

The book is organised into two general chapters (Chapter 1 on marsupial metabolism and nutrient requirements, and Chapter 8 on mineral and vitamin nutrition), together with six chapters devoted to digestion and nutrition in the various groups of marsupials. The species are divided on the basis of their dietary habits into carnivores, omnivores and herbivores. It is among the last mentioned group that we find the most remarkable digestive adaptations. Consequently much more effort has gone into research on the herbivores than the carnivores or omnivores. This book reflects this emphasis in the number of chapters devoted to each group (one each to the carnivores and omnivores, four to the herbivores). Even Chapter 8, on vitamins and minerals, is devoted almost entirely to the herbivores.

Another consequence for this book of the research emphasis on herbivores is that much greater mention is made of Australian marsupials than of American forms; herbivorous marsupials are absent from America. Even among the carnivorous and omnivorous groups, more is known about Australian than American species. However, recently several papers on metabolic rates of American marsupials have appeared, and it is hoped that these will stimulate other studies on the nutrition and metabolism of these species.

Among the Australian species most of the research on marsupial metabolism and nutrition has been concerned with arid-zone species. This is perhaps not so surprising in view of the fact that about two-thirds of the Australian land surface is arid. Of the total of about 125 living marsupial species in Australia, the distributions of 45 species (i.e. 36%) lie within this arid zone (although the distribution of many of these also extends into the semi-arid and more mesic (i.e. higher rainfall) zones as well). Only recently has increased attention been paid to the nutrition and metabolism of marsupials restricted to more mesic environments. It is hoped that this book may stimulate further studies on marsupials from such areas.

The nomenclature of marsupials to species level adopted in this book is based on that of Kirsch & Calaby (1977). Unfortunately this scheme does not include synonyms. To overcome this problem Kirsch (1977b) gives a list of synonyms to supplement Kirsch & Calaby (1977). The reader is referred to this list in any instance of confusion when looking into original sources of information. One synonym which does not appear in this list is the recently recommended use of *Petauroides* in place of *Schoinobates* (McKay, 1980); I have used *Petauroides*.

Preface

In their nomenclature Kirsch & Calaby (1977) do not use subspecies, but Kirsch (1977b) includes subspecies for many species in his list of synonyms. One exception which is of considerable import to the purposes of this book is the *Macropus robustus* group, the Euros and Wallaroos. I have used the subspecific nomenclature of Richardson & Sharman (1976) to distinguish between the Eastern Wallaroo (*M. robustus robustus*) from the Great Dividing Range in Eastern Australia and the Euro (*M. robustus erubescens*) of central and western Australia.

The common names used for Australian species are those recommended by the Vernacular Names Committee of the Australian Mammal Society (Strahan, 1980). The common names used for eutherian species mentioned are taken from Walker (1975). A list of the marsupial species mentioned in the text, with scientific and common names, will be found in the Appendix.

Because of the very recent nature of much of the information contained in this book, I have depended heavily on the work of several present and recent postgraduate students, the data being either unpublished except in theses which are not readily accessible or in manuscripts in the process of publication. For allowing me to use their unpublished results, I thank Stephen Cork, David Dellow, Helen Fletcher, Bill Foley, Robert Inns, Ken Johnson, Stewart Nicol, Robert Prince, Felix Schlager, Andrew Smith, Judith Wake and Rod Wells.

Hugh Tyndale-Biscoe as general editor and Professor W. V. Macfarlane provided many useful comments and suggestions, and John Calaby and Mervyn Griffiths several obscure but valuable references. Frank Knight very kindly provided the silhouettes of each species appearing in the figures. Anna Dooney, Jean Hansford, Betty Harrison and Kathy Santleben shared the typing of the manuscript, and my wife Desley willingly undertook the arduous tasks of proof-reading and reference checking. To all these people my heartfelt thanks.

Armidale   I.D.H.
March 1981

1

Marsupial metabolism and nutrient requirements

In this chapter a number of aspects of the general metabolism of marsupials will be compared with the monotremes on the one hand, and the Eutheria on the other. It is against this general background that requirements of marsupials for energy, protein and water will be discussed in some detail. The digestive strategies and nutrition of the various groups of marsupials will then be dealt with in the next several chapters.

The major aspect of marsupial biology which clearly distinguishes them from all eutherians is their mode of reproduction (Tyndale-Biscoe, 1973). Anatomically it is the relationship of the urinary and genital tracts to each other that exclusively differentiates marsupials from other mammals, not the distinctive external feature, the pouch (marsupium). For instance, the female Echidna (*Tachyglossus aculeatus*) also develops a pouch while suckling (Griffiths, 1978) but is a monotreme. On the other hand, a pouch is not developed at all in the female of the American marsupial genus *Marmosa* (the murine opossums) or the Australian Numbat (*Myrmecobius fasciatus*), and some other marsupial species develop one only while suckling.

Basal metabolism and thermoregulation

The first hint that there may also be differences between marsupials and other mammals in their general physiology was the report by Sutherland (1897) that nine marsupials he examined had a body temperature which averaged 3 °C less than eutherians. He concluded that thermoregulation was less effective in marsupials, and even less effective in monotremes, than in the Eutheria, and this showed that monotremes and marsupials represented stages in the evolution of homeothermy:

> The monotremes are, in consideration solely of their more reptilian anatomy, placed lowest in the scale of mammals. Their low temperature would entirely justify, were justification in any way needed, the position assigned to them next to the reptiles... The next stage in the anatomical classification brings us into the order of the marsupials, and here again we make an upward step in view of a temperature higher, but not so high as that of mammals in general; steadier, but not so steady as is usual in all the remaining orders... It is clear, therefore, that there are grades of temperature, and that the mammals which are classed lowest on anatomical grounds are not only of the lowest temperature, but also of the greatest range, and they are likewise, of all animals, those which are under the strongest and most direct influence of the temperature of the environment.

Soon afterwards, Martin (1903) extended the investigation of the development of homeothermy. He estimated heat production by measuring carbon dioxide production from animals placed in a metabolism chamber, the temperature of which he could vary between 4 and 35 °C, and sometimes up to 40 °C. Martin's observations and conclusions are of considerable importance, as they dominated marsupial physiology for the next 50 years, although sometimes being misinterpreted. His measurements of body temperature confirmed the gradation shown by Sutherland (1897), while in response to a range of environmental temperatures:

> none of the animals maintained a constant body temperature throughout the experimental range. The Marsupials, however, kept their temperature more constant than the rabbit, though less so than the cat.

He concluded that:

> Marsupials show evidence of utilizing variation in (heat) loss to an extent greater than *Ornithorhynchus*, but less than higher Mammals.

Martin's (1903) measurements of resting carbon dioxide production indicated that the minimum metabolic rate of marsupials was only one-third that of eutherians. He suggested, however, that:

> Imperfect homeothermic adjustment cannot explain this relative economy on their part. It is, I believe, largely due to their better insulation, though the fact that their average body temperature is 2° lower than that of Mammals will account for it to a small extent.

No further work of any significance was carried out on marsupial thermoregulation for more than 40 years. During this period it became generally accepted that marsupials were primitive animals, anatomically and in thermoregulatory ability 'below' the Eutheria. Little attention was paid to Martin's (1903) suggestion that the low body temperatures of marsupials did not necessarily infer that their thermoregulatory ability was inferior to placental mammals.

In 1946 Morrison measured body temperatures in three Central American mammals, including two marsupials, *Metachirus nudicaudatus dentaneus* (the Brown Opossum) and *Didelphis marsupialis etenis* (the Eten Opossum) and one eutherian, *Proechimys semipinosis panamensis* (the Spring Rat). He concluded:

> As marsupials, the opossums might be expected to show homoiothermism inferior to that of 'higher' mammals. This was not the case, *Proechimys* being clearly the least competent of the three mammals in this respect.

Notwithstanding these observations, the belief persisted that the monotremes, marsupials and eutherians represented linear stages in the development of homeothermy. Robinson (1954), in a study including two monotremes and six marsupial species, while showing that marsupials were in fact able to vary heat loss as well as heat production, stated that her results showed an evolutionary trend in homeothermy. However, in a later study (Robinson & Morrison, 1957), the idea of a simple ranking of homeothermic ability had virtually disappeared, the authors concluding that while there were considerable differences between the various families with respect to thermoregulation, these were largely due to accompanying differences in size and behaviour and that generally, temperature regulation of marsupials compared well with placental mammals.

Thus clearly where the marsupials differ from the Eutheria is not in ability to regulate body temperature, but in the level at which mean body temperature is set, as hinted by Martin (1903).

Only recently have Martin's (1903) estimates of minimum metabolic rate, or basal metabolic rate (BMR) been re-examined with closer scrutiny. The most comprehensive studies have been those of MacMillen & Nelson (1969) on twelve species of dasyurids, and Dawson & Hulbert (1969, 1970) on eight Australian marsupial species from five different families and ranging in weight from 9 g to 54 kg. The results of these two studies are shown in Tables 1.1 and 1.2 respectively.

In eutherian mammals basal energy metabolism has been shown to vary with the 0.75 power of body weight (Kleiber, 1961) according to the equation $BMR = K \cdot Wt^{0.75}$, and in fact the same average relationship has been found to hold for the whole animal kingdom (Hemmingsen, 1960). In theory, since most metabolic activities occur at surfaces, metabolic rate should increase as the square power, whereas body weight increases as the cube power of body size. Thus to compare the BMR of animals of different body size the discrepancy between surface area increase and volume increase should be accommodated by raising body weight to the two-third power (i.e. $kg^{0.67}$), assuming the animal to be a perfect sphere. However, this is not the case. Empirically the power function which best fits available data from the smallest to the largest mammals has been found to be 0.73 (Brody, 1945) or 0.75 (Kleiber, 1961). By international convention (1965 Conference on Energy Metabolism) the latter exponent has been accepted as the standard for use in comparing energy metabolism between animals of different body size.

Table 1.1. *Basal metabolic rate and body temperature of 12 Dasyuridae*

Species	Body weight (kg)	Basal metabolic rate (W)	Basal metabolic rate ($W \cdot kg^{-0.75}$)	Body temperature (°C)
Planigale maculatus	0.0085	0.0598	2.14	34.8
Sminthopsis crassicaudata	0.0145	0.136	3.25	38.6
Antechinus stuartii	0.0221	0.189	3.30	36.7
Antechinomys laniger	0.0242	0.133	2.17	36.6
Antechinus macdonnellensis	0.0431	0.152	1.61	34.2
Dasycercus cristicauda	0.0888	0.258	1.59	37.7
Dasyuroides byrnei	0.0890	0.433	2.66	37.7
Phascogale tapoatafa	0.157	0.712	2.85	37.4
Dasyurus hallucatus	0.584	1.66	2.48	38.1
Dasyurus viverrinus	0.910	2.29	2.46	36.7
Dasyurus maculatus	1.78	2.98	1.93	36.9
Sarcophilus harrisii	5.05	7.90	2.35	36.8

After MacMillen & Nelson (1969).

When the body weight of the animal is expressed in kg the average value of K for eutherian mammals ranging in size from mice to elephants is 70 if the BMR is expressed in $kcal \cdot kg^{-0.75} \cdot day^{-1}$, 288.8 if the BMR is in $kJ \cdot kg^{-0.75} \cdot day^{-1}$, and 3.34 if the BMR is in the SI (Système International d'Unités) units of $W \cdot kg^{-0.75}$.

In Dawson & Hulbert's (1970) data there is little variation about a mean BMR or standard metabolic rate (SMR) of $2.33\ W \cdot kg^{-0.75}$ (Table 1.2), which is approximately 70% of the eutherian mean. There is more variation in MacMillen & Nelson's (1969) estimates from dasyurids, but the mean BMR of $2.39\ W \cdot kg^{-0.75}$ (Table 1.1) is not significantly different from Dawson & Hulbert's (1970) value. These results therefore support Martin's (1903) conclusion of a low minimum metabolic rate in marsupials in principle, though not in degree.

Variations in basal metabolic rate

Although the results of Dawson & Hulbert (1970) at first suggested that a low BMR was common to all marsupials, and hence was a phylogenetic trait, Hulbert & Dawson (1974a) later found an unusually

Table 1.2 *Basal metabolic rate and body temperature of eight Australian marsupials*

Species (Family)	Body weight (kg)	Basal metabolic rate (W)	Basal metabolic rate ($W \cdot kg^{-0.75}$)	Body temperature (°C)
Sminthopsis crassicaudata (Dasyuridae)	0.0141	0.104	2.54	33.8
Antechinus stuartii (Dasyuridae)	0.0365	0.204	2.14	34.4
Petaurus breviceps (Petauridae)	0.128	0.495	2.31	36.4
Perameles nasuta (Peramelidae)	0.686	1.80	2.39	36.1
Isoodon macrourus (Peramelidae)	0.880	2.20	2.42	34.7
Trichosurus vulpecula (Phalangeridae)	1.98	3.49	2.09	36.2
Macropus eugenii (Macropodidae)	4.80	8.19	2.38	36.4
Macropus rufus (Macropodidae)	32.49	32.4	2.38	35.9

After Dawson & Hulbert (1970).

low BMR in the desert species *Macrotis lagotis* (Greater Bilby), only 58% of that expected on the basis of Kleiber's (1961) equation, compared with 73–76% for two bandicoot species from more mesic habitats. They suggested that the low BMR resulted from a reduced output of thyroxine (an ecological adaptation) superimposed upon a low level of cellular metabolism indicative of all marsupials (a phylogenetic difference).

Other examples of deviations from the marsupial 'mean' BMR have been reported. Two dasyurid species, *Pseudantechinus macdonnellensis* and *Dasycercus cristicauda*, in MacMillen & Nelson's (1969) study exhibited a low BMR, only 47% of the eutherian mean. The BMR of the desert-dwelling Hairy-nosed Wombat (*Lasiorhinus latifrons*) is only 42% of the eutherian mean (Wells, 1978). These and other examples, including *Phalanger maculatus* (Plate 1.1), are listed in Table 1.3.

The only 'deviant' marsupials reported with a BMR higher than the marsupial mean are the dasyurids *Sminthopsis crassicaudata* (97% of the value expected from mass in eutherians) and *Antechinus stuartii* (99%) (MacMillen & Nelson, 1969) and the Water Opossum (*Chironectes minimus*) (95% of the value expected from mass in eutherians (McNab,

Plate 1.1. *Phalanger maculatus*, the Spotted Cuscus, an arboreal folivorous marsupial with a basal metabolic rate lower than the 'marsupial mean'. (Ray Williams.)

Table 1.3. *Basal rates of metabolism of some marsupials which deviate from the general marsupial mean of Dawson & Hulbert (1970)*

Species	Body weight (kg)	Basal metabolic rate (W·kg$^{-0.75}$)	(%)[a]	Body temperature (°C)	Reference
Sminthopsis crassicaudata	0.015	3.25	97	38.6	MacMillen & Nelson (1969)
Antechinus stuartii	0.022	3.30	99	36.7	MacMillen & Nelson (1969)
Chironectes minimus	0.946	3.17	95	35.3	McNab (1978)
Marsupial mean	0.014–32.49	2.33 (2.09–2.54)	70	33.8–36.4	Dawson & Hulbert (1970)
Antechinus macdonnellensis	0.043	1.61	47	34.2	MacMillen & Nelson (1969)
Dasycercus cristicauda	0.089	1.59	47	37.7	MacMillen & Nelson (1969)
Macrotis lagotis	1.011	1.99	58	34.9	Hulbert & Dawson (1974a)
Lasiorhinus latifrons	29.93	1.40	42	33.6	Wells (1978)
Phascolarctos cinereus	3.53–4.83	1.75	52	35.7	Degabriele & Dawson (1979)
Phalanger maculatus	3.09–4.83	2.07	62	34.6	Dawson & Degabriele (1973)

[a] Percentage of predicted value from Kleiber's (1961) equation for eutherians.

1978)). This latter value may be a compensation for high rates of heat loss in a semi-aquatic environment. Similarly, among the Prototheria the BMR of the semi-aquatic Platypus is higher than that of the terrestrial Echidna (*T. aculeatus*) (Grant & Dawson, 1978).

Recent studies among the Eutheria have also brought to light numerous instances of deviations from the Kleiber equation. At the upper extreme of the eutherian range are animals with a BMR 25–100% above the predicted value. These include both Arctic (Casey & Casey, 1979) and tropical species (Scholander *et al.*, 1950; Hildwein & Goffart, 1975). At first a high metabolic rate may seem energetically wasteful, but it may be the result of selection pressures promoting rapid exchange of energy and nutrients between the animal and its environment (Kinnear & Shield, 1975). Thus, a high BMR might be expected to be advantageous to animals faced with a future seasonal energy stress; during favourable periods a high metabolic rate would enable rapid tissue synthesis and energy storage as well as accelerated neonatal development and growth. This is characteristic of Arctic species such as the Caribou (*Rangifer tarandus*) (McEwan, 1970), which has a fasting metabolic rate in summer up to 50% higher than the value predicted from Kleiber's equation. In winter months the metabolic rate of many Arctic species is greatly reduced (Irving, Krog & Monson, 1955), allowing for a slower utilisation of energy reserves, as demonstrated by Aleksiuk & Cowan (1969) in Arctic Beavers (*Castor canadensis*).

At the lower end of the eutherian scale are many desert and temperate species with metabolic rates as much as 30% or more below the predicted value (McNab, 1969; Scholander *et al.*, 1950; Taylor & Sale, 1969).

Thus it is becoming increasingly apparent that Kleiber's (1961) relationship was derived from species representing a single thermoregulatory type: strict homeotherms with body temperatures maintained within about ± 2 °C. For such eutherians the equation allows prediction of BMR from the animal's body weight. However, in heterothermic eutherians (those in which the variation in core temperature, either nychthemerally or seasonally, exceeds ± 2 °C (Bligh & Johnson, 1973)), metabolic rate is a function not only of body weight but also of body temperature. In fact Kinnear & Shield (1975) suggested that most of the differences seen between marsupials and eutherians in BMR and body temperature are a result of comparing only rigidly homeothermic eutherians with both heterothermic and homeothermic marsupials. In addition, they suggested that BMR and body temperature were primarily adaptive to the environment, and that comparisons between marsupials and eutherians cannot be indicative of phylogeny.

More recently Nicol (1978a) made a more detailed analysis of the energetics of 30 marsupial and 34 non-domesticated eutherian species. The statistical analysis incorporated all the marsupial species for which complete data were available, and a broad cross-section of eutherians. Analysis of covariance of the eutherian and marsupial regression lines of log BMR against log body weight showed that the two slopes were virtually identical, and indistinguishable from the slope of 0.75 proposed by Kleiber (1961). However, the eutherian line, due to the inclusion of a number of heterothermic species, was some 16% lower than the lines obtained by Kleiber (1932) and Brody & Procter (1932). Nevertheless, the difference between the Y intercepts was still highly significant, and on average, the BMR of a marsupial would be predicted to be about 22% lower than that of a eutherian of the same weight.

Thus the lower BMR of marsupials does not appear to be due entirely to the fact that previous comparisons were made only with 'rigid homeotherms' on the one hand and both heterotherms and homeotherms on the other as Kinnear & Shield (1975) asserted.

The other way in which the two groups differed in Nicol's (1978a) analysis was in the relationship between BMR and body temperature (Fig. 1.1). This suggested that while a large part of the differences in basal metabolism between many eutherian species can be associated with differences in body temperature, this is not the case with the 30 marsupial species analysed, since the slope of the marsupial regression was not significantly different from zero.

Despite differences between the two groups in the relationship between BMR and body temperature, overall the difference in body temperature between marsupials and eutherians was not statistically significant. This would seem to disprove the assertion which is sometimes made that the differences in BMR between the two groups is at least partly explicable in terms of differences in body temperature (e.g. Kinnear & Shield, 1975).

In Nicol's (1978a) view, then, BMR and body temperature are governed more by phylogeny than by habitat. However, this is not to say that within comparable habitats there will always be differences between marsupial and eutherian metabolic rates, or that within certain groups of eutherians BMR will always be directly related to body temperature. In biology boundaries are rarely clearly defined, and exceptions can often be found to general overall relationships such as those just discussed. An exception to the close relationship between BMR and body temperature among eutherians is found in a study of hedgehogs by Shkolnik & Schmidt-Nielsen (1976) in which three species had similar body temperatures, yet BMR decreased with increasing aridity of the habitat. Similarly, the overall difference in BMR between marsupials and eutherians disappears when species of both groups occupying tropical arboreal habitats are compared, and also those occupying arid fossorial habitats.

McNab (1978) has taken the comparison of eutherians and marsupials in comparable habitats further by categorising animals into six groups (Table 1.4). The first, aquatic carnivores, contains too few numbers of

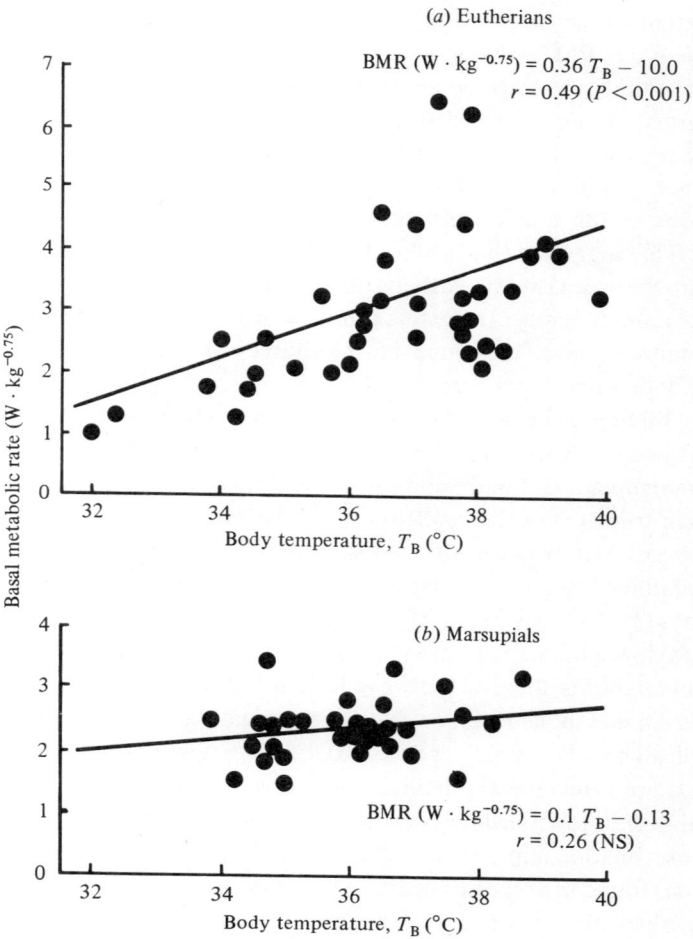

Fig. 1.1. The relationship between basal metabolic rate and body temperature of 30 marsupial and 34 non-domesticated eutherian species. Part of the differences in basal metabolism between eutherian species, but not marsupial species, is associated with differences in body temperature. After Nicol (1978a).

(a) Eutherians

$BMR (W \cdot kg^{-0.75}) = 0.36 \, T_B - 10.0$
$r = 0.49 \, (P < 0.001)$

(b) Marsupials

$BMR (W \cdot kg^{-0.75}) = 0.1 \, T_B - 0.13$
$r = 0.26 \, (NS)$

either marsupials or eutherians to allow realistic comparisons to be made. Similarly the last group, terrestrial omnivore-insectivores, contains only one marsupial group, the bandicoots. Of the remaining four classifications, terrestrial marsupials all have a BMR below the eutherians, but the difference disappears in the two arboreal groups, as Nicol (1978a) found.

In McNab's (1978) view, the BMR of marsupials is primarily correlated with food habits, although there may be some (modest) influence of climate. However, McNab (1978) added the cautionary note that a correlation between BMR and food habits or climate cannot establish,

Table 1.4. *Basal rates of metabolism in marsupials and their placental analogues*

	Body weight (kg)	Basal metabolic rate	
		% of eutherian standard[a]	$\frac{\text{Marsupial}}{\text{Eutherian}} \times 100(\%)$
A. *Aquatic carnivores*			
Marsupial (*Chironectes minimus*)	0.946	95	82
Eutherian (*Mustela vison*)	0.66, 0.70	110, 122	
B. *Terrestrial carnivores*			
Marsupial mean (4)[b]	0.09–0.91	76.3 ± 1.6	57
Eutherian mean (3)	0.15–0.30	132.7 ± 16.2	
C. *Terrestrial grazer-browsers*			
Marsupial mean (5)	0.95–32.49	69.6 ± 1.1	57
Eutherian mean (3)	2.30–100.0	126.3 ± 6.8	
D. *Arboreal folivores*			
Marsupial mean (3)	0.86–4.25	59.3 ± 3.2	111
Eutherian mean (5)	2.21–4.25	53.2 ± 5.2	
E. *Arboreal frugivore–omnivores*			
Marsupial mean (8)	0.07–3.26	74.1 ± 1.7	106
Eutherian mean (4)	0.07–2.03	69.5 ± 3.5	
F. *Terrestrial omnivore–insectivores*			
Marsupial mean (5)	0.65–0.88	69.4 ± 2.9	144
Eutherian mean (4)	1.11–2.15	47.5 ± 2.7	

After McNab (1978).
[a] From Kleiber's (1961) equation.
[b] Parentheses: number of species examined.

per se, a causative relationship. A great deal more information is needed on the BMR of mammals, both marsupial and eutherian, from different habitats and of different feeding types, before much more can be said about McNab's (1978) hypothesis.

In summary, there are numerous examples of deviations in BMR from the 'Kleiber mean' among eutherians, and from the 'Dawson and Hulbert mean' among marsupials. Most of these deviations in both groups can be correlated with habitat, including climate and/or food habits. However, even when deviations from the respective means are included, marsupials have basal metabolic rates covering only the lower portion of the eutherian range, and overall the difference between marsupials and eutherians is statistically significant. This is a modification of the conclusion of Tyndale-Biscoe (1973) that a low metabolic rate appears to distinguish the Marsupialia from the Eutheria as clearly as anatomical criteria. This reflects the continuing progress that has been achieved in marsupial physiology in the last ten years, and the value of comparative studies in animal biology.

The consequences of a low metabolic rate

A low metabolic rate has several important consequences for animals in terms of nutrient requirements and environmental tolerance. For instance, we can confidently assume that a low BMR will mean lower food requirements, and energy reserves in the body will last longer under adverse conditions. Thus, within the Eutheria, a relatively low BMR appears to be a common characteristic among desert rodents (Dawson, 1955; McNab & Morrison, 1963). The remainder of this chapter will be devoted to a discussion of some of the consequences of a low BMR for marsupials.

Kinnear & Brown (1967) and Dawson & Bennett (1978) found a lower heart rate in marsupials than in eutherians. The minimum heart rates of 14 species of marsupials were a little less than half those of eutherian species of comparable body weight (Kinnear & Brown, 1967). The regression equation relating minimum heart rate (F) to body size in kg (W) for the marsupials was $F = 106\ W^{-0.27}$ compared with the equation $F = 217.8\ W^{-0.27}$ given by Brody (1945) for mature eutherians. Since the haemodynamics of marsupials appear to be comparable to those of eutherians (Maxwell, Elliott & Kneebone, 1964), the low heart rates in marsupials are taken to reflect a lower level of energy metabolism compared with most placental species.

In desert rodents a relatively low BMR is associated with low circulatory thyroxine levels (Scott, Yousef & Johnson, 1976). This has prompted several workers to examine thyroid function in marsupials. Setchell (1974) found a mean plasma total thyroxine concentration of 1.6 ± 0.1 $\mu g \cdot dl^{-1}$ for five macropodid species, which was significantly lower than the mean value he obtained for three eutherians (4.3 ± 0.4 $\mu g \cdot dl^{-1}$). Nicol (1978a) found similarly low total thyroxine concentrations in the plasma of the Potoroo (*Potorous tridactylus*) but the concentration of free thyroxine, the active form of the hormone, was within the eutherian range. In agreement with this finding, Setchell (1974) demonstrated that the rates of thyroid secretion were similar for marsupials and eutherians of the same size despite their different metabolic rates; thus marsupials appear to use 30% more thyroxine per unit of metabolism. He concluded that any difference in metabolic rate between these two groups of mammals is independent of thyroid hormone secretion, and that the level of metabolism is determined genetically. Similarly, within the Artiodactyla, Macfarlane (1977) found no relationship between thyroxine turnover and metabolic rate; cattle and camels in the same environment had similar thyroxine turnovers but widely different metabolic rates.

The role of the thyroid hormones thus appears to be to modulate

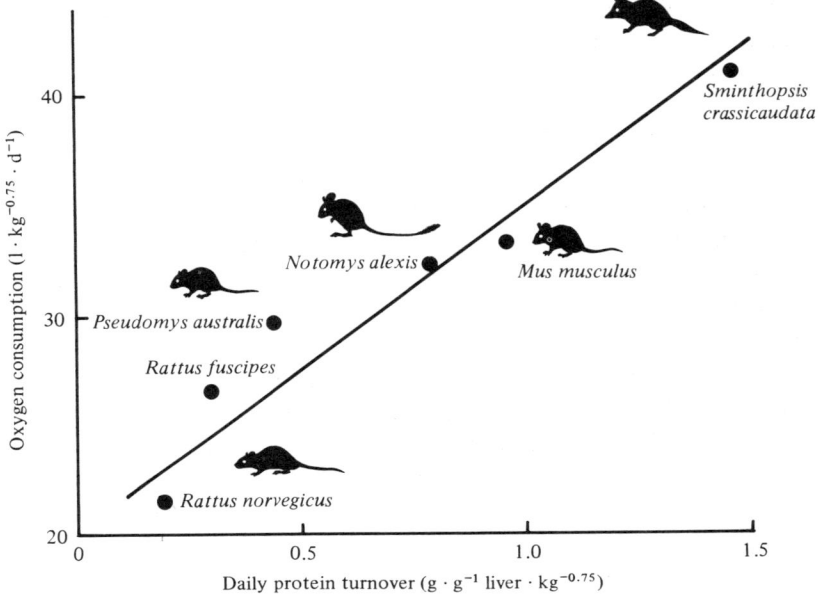

Fig. 1.2. The relationship between basal metabolic rate (minimum rate of oxygen consumption) and liver protein turnover in *Sminthopsis crassicaudata* and five rodents. The rate of protein synthesis is probably a major component of basal metabolic rate. From data of Macfarlane (1977).

metabolic rate about a predetermined level in response to seasonal environmental conditions, primarily temperature, as Kaethner & Good (1975) showed for the Tammar Wallaby (*Macropus eugenii*) and Bauman & Turner (1966) found for the American Opossum (*Didelphis virginiana*). The other factor which influences thyroxine secretion rate is food intake; thyroxine is produced to meet the needs of metabolising energy taken in as seasonal food supplies change (Macfarlane, 1977).

According to Macfarlane (1976), the predetermined level of basal metabolism is most probably determined by the inherent rate of protein synthesis in a species. Since 40 kJ of energy is needed to synthesise one gram of protein, the rate of synthesis of proteins is probably a major component of BMR. This is suggested by the relationship shown in Fig. 1.2 for *Sminthopsis crassicaudata* and five rodent species.

Energy requirements

To maintain itself in longer term energy balance, an animal needs energy in addition to that for basal metabolism; that is, for activity associated with feeding and drinking, and for thermoregulation. Estimates of the total energy cost of maintenance, or free existence, vary between 1.68 times the BMR for a sitting (resting) tropical hummingbird (*Eulampis jugularis*) at the field temperature (Wolf & Hainsworth, 1971), 1.7 to 2.4 times the BMR for Purple Martins (*Progne subis*) and Mockingbirds (*Mimus polyglottos*) that were not actively feeding nestlings, and three times the BMR of average weight individuals of 14 species of large mammals in the Tarangirie Game Reserve in East Africa (Lamprey, 1964). For most domesticated herbivores maintenance energy turnover is approximately twice the BMR (Maynard & Loosli, 1962). Similarly, Nagy & Milton (1979) estimated that field energy expenditure by wild Howler Monkeys was about twice their calculated BMR.

If the same relationship holds for marsupials, their maintenance energy requirements should also be less than those of eutherians. This indeed appears to be so. Hume (1974) determined the maintenance energy requirements of the Euro (*Macropus robustus erubescens*), the Red Kangaroo (*M. rufus*) and sheep by feeding them roughages of three different qualities in the laboratory. Nitrogen balance was measured and then plotted against digestible energy intake for each species on a metabolic body weight basis. On the assumption that in adult animals zero nitrogen balance coincides with zero energy balance, the maintenance energy requirement of the sheep was calculated to be 136 kcal digestible energy $\cdot$ kg$^{-0.75}\cdot$ day^{-1}, or 6.49 W $\cdot$ kg$^{-0.75}$. Note that this is almost double the BMR of 70 kcal·kg$^{-0.75}$·day^{-1} or 3.34 W·kg$^{-0.75}$ established for homeothermic eutherians by Kleiber (1961). In contrast the estimated maintenance requirement of the Red Kangaroo was 109 kcal·kg$^{-0.75}$·day^{-1} (5.20 W·kg$^{-0.75}$), and of the Euro 99 kcal·kg$^{-0.75}$·day^{-1} (4.72 W·kg$^{-0.75}$), 80% and 73% respectively, of the requirement for sheep. Again these maintenance requirements are approximately double the BMR.

The only other estimates of maintenance energy requirements of marsupials are of dasyurids. Green & Eberhard (1979) estimated that the digestible energy required for maintenance of captive Tasmanian Devils (*Sarcophilus harrisii*) and Eastern Quolls (*Dasyurus viverrinus*) was 545 kJ·kg$^{-0.75}$·day^{-1} or 6.30 W·kg$^{-0.75}$. This is higher than the values for maintenance requirement of the Euros and Red Kangaroos of Hume (1974), approximately 2.5 times their calculated BMR. Nevertheless, the maintenance energy requirements of eutherian carnivores such as the Mink (*Mustela vison*) (Farrell & Wood, 1968), and the Red Fox (*Vulpes fulva*) (Vogtsberger & Barrett, 1973), in captivity are significantly higher than those of the two marsupial carnivores studied by Green & Eberhard (1979).

Higher estimates of energy expenditure have been reported by Cowan, O'Riordan & Cowan (1974) in the alpine dasyurid *Antechinus swainsonii* maintained in cages for eight weeks. At weight maintenance their digestible energy consumption was close to four times the calculated BMR. The authors interpreted the high maintenance estimate as representing the energy cost of maintenance plus activity, since the animals were extremely active in their cages, and concluded that this total energy expenditure was probably close to the normal energy demand of this species under free-living conditions. Nagy, Seymour, Lee & Braithwaite (1978) reported similarly high energy expenditures in free-living *Antechinus stuartii*, using the doubly labelled water (^{3}H$_2$^{18}O) field technique of Lifson & McClintock (1966) in association with laboratory feeding trials. Total energy turnover was about 4.5 times their basal rate (MacMillen & Nelson, 1969), but ambient temperatures averaged only 8 °C during the mid-winter measurement period, and it was calculated that about 85% of their energy expenditure would have been allocated to basal metabolism and thermoregulation. Nevertheless, as with the results from *A. swainsonii*, the present estimates are probably a realistic reflection of the total energy requirements of free-living *A. stuartii* at that time of the year.

Smith (1980) also found a large difference between field energetic requirements of the omnivorous arboreal species *Gymnobelideus leadbeateri* (Leadbeater's Possum) and its BMR. Resting metabolic rates (BMR) in captive animals were 92% of that predicted for a marsupial on the basis of bodyweight by Dawson & Hulbert (1970). However, field energy requirements were 4.8 times BMR. By monitoring the activity of one colony of *G. leadbeateri* Smith (1981) was able to partition total energy expenditure in spring into the cost of activity outside the nest (70%), the cost of thermoregulation outside the nest (16%) and the cost of resting inside the nest (14% of average daily energy requirements). The high contribution of activity outside the nest to total energy expenditure was consistent with the observation that when outside Leadbeater's Possums are extremely active and fast moving, and are rarely seen to be immobile like many of the larger folivorous possum species.

Nagy & Suckling (in Smith, 1980) also found that field energy requirements of the arboreal Sugar Glider (*Petaurus breviceps*) were 4.5 times their BMR in spring under environmental conditions similar to those experienced by *G. leadbeateri* in Smith's (1980) study.

Thus, as has been shown with eutherians, the total energy requirements of free-living marsupials will exceed the maintenance values of captive animals by an extent depending upon the cost of additional requirements for activity and thermoregulation. The additional energy costs for growth and reproduction are likely to be even larger, as they have been found in birds (West, 1960) and in placental mammals (Maynard & Loosli, 1962), but no information has yet been published for marsupials.

The lower BMR of marsupials, and the lower maintenance energy

Table 1.5. *Voluntary feed intake of seven macropodine species and of sheep fed chopped lucerne hay. All values given as g dry matter·$kg^{-0.75}$·day^{-1}; number of animals in parentheses*

Sheep	Macropodine	Reference
71.7 (3)	58.1 (3) *Macropus rufus*	Foot & Romberg (1965)
64.1 (4)	38.7 (4) *Macropus rufus*	McIntosh (1966)
66.5 (3)	63.4 (1) *Macropus rufus*	Forbes &
	48.1 (4) *Macropus giganteus*	Tribe (1970)
79.0 (2)	53.0 (2) *Macropus giganteus*	Kempton (1972)
91.6 (3)	53.4 (3) *Macropus rufus*	Hume (1974)
	52.7 (3) *Macropus robustus*	
62.0 (4)	54.6 (4) *Macropus rufogriseus*	Hume (1977a)
	69.2 (4) *Thylogale thetis*	
60.3 (7)	56.7 (8) *Macropus giganteus*	Dellow (1979)
	52.7 (9) *Thylogale thetis*	
	29.4 (8) *Macropus eugenii*	
	47.3 (4) *Setonix brachyurus*	Calaby (1958)

requirements of captive macropodine marsupials compared with eutherians, is directly reflected in the voluntary feed intakes of adult animals at or near body weight maintenance. This is illustrated in Table 1.5, which includes data from seven studies in which macropods and sheep were fed the same chopped lucerne hay diet under laboratory conditions. Data for the Quokka, *Setonix brachyurus*, are also included, although sheep were not compared directly in Calaby's (1958) study. In the study of Hume (1977a) the dry matter intakes of Red-necked Pademelons (*Thylogale thetis*) were significantly higher than those of the sheep. In all other cases the voluntary intake of the macropodines was lower, the lowest being that of the Tammar Wallaby (*Macropus eugenii*). The seemingly anomalous results for *T. thetis* will be referred to again in relation to requirements for nutrients other than energy.

Nitrogen requirements

The total requirement for nitrogen is much less affected by additional requirements for free existence, such as activity and thermoregulation, than is energy. Thus estimates of the maintenance requirement of captive animals for nitrogen are likely to be a realistic reflection of the needs of adult animals in the wild. Only growth and reproduction impose significant increments on the total protein requirement, but, as with energy, protein requirements for these physiological functions in marsupials have not been investigated.

If, as suggested by Macfarlane (1976), the main determinant of the level of basal metabolism is the rate of protein synthesis in the animal, there should be a close relationship between some measure of protein turnover and basal energy turnover. Indeed, in eutherians Smuts (1935) has shown that about 2 mg of endogenous urinary nitrogen is excreted for each kcal of basal heat production in all species investigated. If the same relationship holds for marsupials, the endogenous nitrogen losses, and the maintenance nitrogen requirements of marsupials, should be lower than eutherians. Brown (1968) and Barker (1968) presented data to support the notion of low endogenous urinary nitrogen losses in at least five macropodine species; these are compared with eutherian values in Table 1.6. Note that the eutherian mean of Brody (1945) of 146 mg $N \cdot kg^{-0.75} \cdot day^{-1}$ is essentially twice the mean eutherian BMR of 70 $kcal \cdot kg^{-0.75} \cdot day^{-1}$. The lowest value was obtained for the heterothermic camel, which has a BMR below the homeothermic mean (Schmidt-Nielson *et al.*, 1967). The data from the macropods are more variable, but the mean value of about 55 mg $N \cdot kg^{-0.75} \cdot day^{-1}$ is well below the eutherian mean.

In most marsupial species studied this low mean endogenous urinary output can be directly related to a low maintenance requirement for nitrogen. Brown & Main (1967) were the first to publish such estimates. These and subsequent estimates for other marsupial species are shown in Table 1.7, along with estimates for some eutherians.

In this instance there is a greater number of reliable estimates for marsupials than for eutherians. Among the Eutheria the Rock Hyrax (*Procavia habassinica*) is distinctive in that its maintenance nitrogen requirement lies well below that of most other eutherians studied, and is much more similar to that of most marsupials; its BMR is also low (Taylor & Sale, 1969). Among the marsupials the Red-necked Pademelon is more similar in its maintenance nitrogen requirement to the sheep rather than

Table 1.6. *Minimum or endogenous urinary nitrogen excretion of macropods and eutherians*

Species	Body weight (kg)	Minimum urinary nitrogen excretion (mg·kg$^{-0.73}$·day^{-1})	Reference
Macropus robustus erubescens (Euro)	8.5–19.7	34 (21–48)	Brown (1968)
Macropus rufus (Red Kangaroo)	14.4[a]	92	Brown (1968)
Petrogale sp. (Rock-wallaby)	3.9[a]	50	Brown (1968)
Setonix brachyurus (Quokka)	2.0–3.6	44	Brown (1968)
Macropus eugenii (Tammar)	4.2–5.3	60 (40–80)	Barker (1968)
Macropod mean		55	
Rat	0.17–0.23	137 (111–157)	Brody (1945)
Guinea pig	0.33–0.50	140 (129–154)	Brody (1945)
Rabbit	1.16–2.78	150 (116–175)	Brody (1945)
Pig	24.0–79.0	131 (109–141)	Du Toit & Smuts (1941)
Goat	24.2–62.0	124 (112–137)	Hutchinson & Morris (1936)
Sheep	31.8–42.0	94 (78–104)	Smuts & Marais (1938)
Camel	250	67	Schmidt-Nielsen et al. (1957)
Eutherian mean		146	Brody (1945)

[a] Only one animal.

Table 1.7. *Maintenance nitrogen requirements of marsupials and eutherians; all data in mg* $N \cdot kg^{-0.75} \cdot day^{-1}$

Species	Body weight (kg)	Maintenance requirement		Reference
		Dietary	Truly digestible	
Macropus robustus erubescens (Euro)	10.5–15.6	290	160	Brown & Main (1967)
M. eugenii (Tammar Wallaby)	4.2–5.4	290	250	Barker (1968)
	3.8–5.1	240	230	Hume (1977b)
M. robustus robustus (Eastern Wallaroo)	11.8–17.7	300	240	Foley et al. (1980)
M. giganteus (Eastern Grey Kangaroo)	18.6–30.3	350	270	Foley et al. (1980)
Thylogale thetis (Red-necked Pademelon)	3.8–4.8	600	530	Hume (1977b)
Trichosurus vulpecula (Brushtail Possum)	1.3–2.5	203	189	Wellard & Hume (1981a)
Phascolarctos cinereus (Koala)	5.1–8.3	283	270	Cork (1981)
Lasiorhinus latifrons (Hairy-nosed Wombat)	29.0–31.0	205	—	Wells (1968)
Ovis aries (Sheep)	31.8–42.7	489	452	Moir & Williams (1950)
		390	340	Harris & Mitchell (1941)
Equus caballus (Horse)	119	—	380	Prior et al. (1974)
Procavia habessinica (Rock Hyrax)	2.0–2.6	311	209	Hume et al. (1980)

to the other marsupials studied. Notwithstanding these exceptions, generally the maintenance nitrogen requirements of marsupials appear to be lower than estimates from eutherians, in line with their generally lower maintenance requirement for energy and their lower average BMR.

The high nitrogen requirement of the Red-necked Pademelon suggests that either its BMR is higher than the marsupial mean of Dawson & Hulbert (1970), or the relationship between endogenous urinary nitrogen and basal energy metabolism may differ between marsupial species. Present evidence indicates no significant variation in BMR among the macropodids. However, the BMR of *Thylogale* species has not been measured directly. The relatively high voluntary feed intakes of adult Pademelons at or near body weight maintenance (Table 1.5) suggest that the forest-dwelling *Thylogale thetis* may have a maintenance energy requirement, and perhaps also a BMR, higher than the macropodid mean. However, the only other forest-dwelling macropodid species studied in this context, the Potoroo (*Potorous tridactylus*) has a BMR similar to the arid and semi-arid zone species (Hudson & Dawson, 1975; Nicol, 1976).

The other possibility, that the relationship between endogenous urinary nitrogen and basal energy metabolism may differ among marsupial species, cannot be fully tested with data presently available. This is partly due to the limited number of species examined, and partly to the variability in current estimates of excretion of creatinine, a commonly used measure of endogenous nitrogen metabolism (Mitchell, 1962). Creatinine is a metabolic end-product of creatine, a precursor of the high-energy compound phosphocreatine found in muscle. Creatine is synthesised in the liver. Surplus creatine from the reversible interconversion between creatine and phosphocreatine is converted to creatinine, which is excreted by the kidney. The rate of creatinine excretion by a healthy animal fed a diet free of creatine and creatinine appears to be equivalent to the rate of creatine synthesis in the liver, which proceeds at a rate proportional to BMR. Unlike urea, another end product of nitrogen metabolism, creatinine does not appear to enter the digestive tract to be degraded by microorganisms. Creatinine excretion does not seem to be affected by muscular activity or other factors that increase metabolic rate, although it is disturbed by any factor that raises deep body temperature (Mitchell, 1962).

Creatinine excretion has been measured in five macropodid species, and the results are included in Table 1.8. As with minimum urinary nitrogen excretion (Table 1.6), the macropodid mean is below that of the eutherians. Among the macropodids, although standard deviations are large in some cases, it does appear that there may be interspecific differences in creatinine excretion. Of particular interest is the fact that the Euro is consistent in having the lowest values for both creatinine excretion (Table 1.8) and minimum nitrogen excretion (Table 1.6), as well as maintenance nitrogen requirements (Table 1.7). Of further interest is the Red-necked Pademelon, which, along with its unusually high maintenance nitrogen requirement, has one of the highest values for creatinine excretion. Whether the differences among the macropodid species in creatinine excretion can be related to differences in BMR, though, is unknown. What is required is more precise information on creatinine excretion, as well as careful estimation of BMR, in a wide range of marsupial species, especially those from mesic environments which have received scant attention in the past.

Water turnover

It has often been stated that the rate of water turnover of an animal is related to its metabolic rate (e.g. Hudson, 1962; Macfarlane et al., 1971). Thus it is of interest to know whether marsupials differ from eutherians in their water turnover rates in the same way that they differ in their BMR. Since water turnover is likely to be influenced by several factors, including activity, climate and access to water (Hulbert & Gordon, 1972), any

Table 1.8. *Creatinine excretion in five macropodid species and five eutherian species*

Species	Body weight (kg)	Creatine excretion ($mg \cdot kg^{-0.75} \cdot day^{-1}$)	Reference
Potorous tridactylus (Long-nosed Potoroo)	1.12	40.9 ± 15.6 (SD)	Nicol (1976)
Setonix brachyurus (Quokka)	3.25	34.1 ± 10.9 (SD)	Ramsay (1966)
Macropus robustus erubescens (Euro)	14.63	24.0 ± 5.4 (SD)	Fraser & Kinnear (1969)
Macropus eugenii (Tammar)	3.79	30.2 ± 6.0 (SD)	Fraser & Kinnear (1969)
	2.4–5.9	29.2 (25.8–33.1)	Kinnear & Main (1975)
	4.3 ± 0.3	23.7 ± 7.9 (SE)	Wilkinson (1979)
Thylogale thetis (Red-necked Pademelon)	4.3 ± 0.2	39.1 ± 16.1 (SE)	Wilkinson (1979)
Ovis aries (Sheep)	37–50	62.2	Fraser & Kinnear (1969)
Sus scrofa (Pig)	24–79	55.7	Smuts (1935)
Oryctolagus cuniculus (Rabbit)	2.0	55.3	Brody (1945)
Camelus dromedarius (Camel)	515.0	85.3	Brody (1945)
Bos taurus (Cattle)	322.0	110.4	Brody (1945)

comparisons should be made under standardised conditions, just as BMR must be measured under truly basal conditions. Nicol (1978b) has suggested that the ambient temperature used should be at the lower end of the thermoneutral zone, since a higher temperature may result in increased water loss for evaporative cooling, while a lower temperature will raise the metabolic rate and thus increase water turnover. Water must be available *ad libitum*, since water deprivation reduces metabolic rate (Schmidt-Nielsen *et al.*, 1967) and water turnover (Hulbert & Dawson, 1974b; Nagy, Shoemaker & Costa, 1976). Alternatively, food containing a high proportion of water will supply adequate amounts (Rübsamen, Heller, Lawrenz & Engelhardt, 1979a). Water turnover measured under conditions of adequate water availability and an ambient temperature at the lower end of the thermoneutral zone might then be described as the standard water turnover rate (SWTR) (Nicol, 1978b).

Water turnover rate can be estimated from the dilution rate of a single dose of tritiated water (THO) in any body fluid; either blood or urine can be used (Rübsamen, Nolda & Engelhardt, 1979b). If evaporative water is used there will be a small but significant error introduced by differential movements of hydrogen and tritium across membranes (Rübsamen *et al.*, 1979b).

Using tritiated water, Richmond, Langham & Trujillo (1962) found that the mean water turnover rate in seven species of eutherians ranging in size from the 21 g house mouse to the 399 kg horse, with water available *ad libitum*, was 134 ± 32 ml·kg$^{-0.80}$·day^{-1}. Denny & Dawson (1975a) subsequently showed that the mean water turnover rate in five macropodid species held under laboratory conditions, again with water available *ad libitum*, was 98 ± 21 ml·kg$^{-0.80}$·day^{-1}. Although variation about the mean in both studies is considerable, the macropodid mean is 27% less than the eutherian mean. Other measurements by Denny & Dawson (1973), Kennedy & Heinsohn (1974), Haines, Macfarlane, Setchell & Howard (1974) and Hulbert & Dawson (1974b) support the concept of a generally low standard water turnover rate in marsupials. These and other results are shown in Table 1.9.

A low standard water turnover rate in marsupials in addition to a low BMR is not entirely unexpected, since the various avenues of water loss from the body are all indirectly influenced by the animals' metabolic rate. For example, faecal water loss is determined in part by the amount of food eaten, which we have seen to be lower in most macropodid species than in most eutherians (Table 1.5). Urine water loss is partly determined by glomerular filtration rate, which is lower in kangaroos than in most other

Table 1.9. *Total body water content and water turnover rates of marsupials and eutherians*

Species	Body weight (kg)	Total body water (% of body weight)	Water turnover (ml·kg$^{-0.80}$·day^{-1})	Reference
Macropus giganteus	22.1 ± 1.5 (6)	78.0 ± 2.5	78.5 ± 5.8	Denny & Dawson (1975a)
M. robustus robustus	31.1 ± 5.2 (5)	77.6 ± 2.7	117.9 ± 17.2	Denny & Dawson (1975a)
M. rufus	23.4 ± 1.6 (6)	72.5 ± 2.3	119.7 ± 10.9	Denny & Dawson (1975a)
M. eugenii	6.5 ± 0.2 (6)	60.8 ± 2.0	65.2 ± 5.0	Denny & Dawson (1975a)
Potorous tridactylus	1.4 ± 0.1 (5)	58.8 ± 2.3	105.9 ± 4.0	Denny & Dawson (1975a)
M. rubustus erubescens	24.1 ± 3.4 (13)	74.3 ± 1.5	95.6 ± 12.3	Denny & Dawson (1973)
Petrogale inornata	3.2 ± 0.3 (8)	68.2 ± 0.5	98.9 ± 2.2	Kennedy & Heinsohn (1974)
Lasiorhinus latifrons	25 (5)	72.2 ± 2.2	33.4 ± 9.5	Wells (1973)
Macrotis lagotis	1.1 ± 0.2 (4)	56.6 ± 2.7	45.6 ± 2.6	Hulbert & Dawson (1974b)
Perameles nasuta	1.0 ± 0.1 (6)	61.9 ± 0.9	72.5 ± 5.7	Hulbert & Dawson (1974b)
Isodon macrourus	1.5 ± 0.2 (5)	56.7 ± 2.1	97.5 ± 8.1	Hulbert & Dawson (1974b)
Dasyuroides byrnei	0.13 ± 0.01 (5)	67.7	97.8 ± 7.9	Haines et al. (1974)
Dasycercus cristicauda	0.086 (34)	56–68	95.1 ± 3.3	Kennedy & Macfarlane (1971)
Sminthopsis crassicaudata	0.019 (8)	ND	237.8 ± 12.2	Kennedy & Macfarlane (1971)
Trichosurus vulpecula	1.52 ± 0.03 (31)	68.6 ± 1.8	96.0 ± 1.7	Kennedy and Heinsohn (1974)
Sminthopsis crassicaudata	0.015 ± 0.003 (12)	72.9 ± 9.0	217.0 ± 38.0	Morton (1980a)
Mus musculus	0.021	58.5 ± 4.0	157.0 ± 25.0	Richmond et al. (1962)
Rattus norvegicus	0.298	59.6 ± 4.0	90.9 ± 12.9	Richmond et al. (1962)
Oryctolagus cuniculus	3.159	58.4 ± 5.3	134.7 ± 24.7	Richmond et al. (1962)
Canis domesticus	10.58	66.0 ± 1.4	143.3 ± 18.8	Richmond et al. (1962)
Homo sapiens	67.30	55.3 ± 5.3	94.7 ± 17.9	Richmond et al. (1962)
Equus caballus	398.5	65.7 ± 0.7	180.6 ± 27.0	Richmond et al. (1962)

ND: not determined.

mammals studied (Denny & Dawson, 1975a). Insensible water loss via the skin would tend to be low because of the lower body temperatures in marsupials (Dawson & Hulbert, 1970), and respiratory water loss would also tend to be reduced due to the lower oxygen consumption in marsupials (Dawson, Denny & Hulbert, 1969).

However, as with energy turnover, low water turnover rates have also been recorded in some species of eutherians, primarily desert-adapted species. One of the lowest is 69 ml·kg$^{-0.80}$·day^{-1} recorded for the desert rodent *Dipodomys merriami* (Merriam's Kangaroo-rat) by Yousef, Johnson, Bradley & Seif (1974). Similarly, among the marsupials water turnover rates as low as 33 ml·kg$^{-0.80}$·day^{-1} have been reported, in this case for the desert-dwelling Hairy-nosed Wombat (*Lasiorhinus latifrons*) (Wells, 1973). At the other end of the scale Holleman & Dieterich (1973) reported a water turnover rate of 194 ml·kg$^{-0.80}$·day^{-1} in *Microtus oeconomus* (the Tundra Vole), but with the exception of *Sminthopsis crassicaudata* (238 ml·kg$^{-0.80}$·day^{-1} (Kennedy & Macfarlane, 1971)), such high rates under standard conditions among the marsupials have not been reported. Kennedy & Macfarlane's (1971) value for *S. crassicaudata* (see Plate 1.2) has been discounted by Hulbert & Dawson (1974b) as being too high, as

Plate 1.2. *Sminthopsis crassicaudata*, the Fat-tailed Dunnart, a small carnivorous marsupial with unusually high rates of metabolism and water turnover. (David Walsh.)

the associated oxygen consumption value was much higher than Dawson & Hulbert's (1970) value for this species; they argued that measurement conditions did not appear to be basal. However, Kennedy & Macfarlane's (1971) high value for water turnover rate has now been substantiated by Morton (1980a). Inconsistent with this is Morton's (1980a) value of BMR in *S. crassicaudata* of 2.54 $W \cdot kg^{-0.75}$ which is exactly the same as that of Dawson & Hulbert (1970), and thus less than Kennedy & Macfarlane's (1971) value of 3.92 and also MacMillen & Nelson's (1969) value of 3.23 $W \cdot kg^{-0.75}$. This range in BMR may be a reflection of the readiness with which *S. crassicaudata* enters torpor, with consequent very low oxygen consumption rates (Godfrey, 1968). The use of torpor and other measures by the small dasyurids to reduce their total turnover of energy and water will be discussed in greater detail in the next chapter.

Macfarlane (1965) was the first to suggest that water turnover rates of animals measured with water freely available can be used to separate 'desert hardy' species from others. That is, animals with a high water turnover under 'non-stress' or standard conditions were likely to be those that were found in water-rich habitats. Denny & Dawson's (1973, 1975b) studies indicated no correlation between *ad libitum* water turnover rate and aridity of the present habitat of six macropodid species, even though habitats ranged from arid (for the Euro) to moist forest (Potoroo). They were thus unable to concur with Macfarlane's (1965) proposal, and concluded that any study of an animal's ability to survive in an arid area should include work done on the animal under conditions similar to those of the arid zone.

More recently, however, Nicol (1978b) has analysed standard water turnover rate data for 13 species of marsupials and 27 eutherian species in a way similar to that used in his analysis of the BMR of the two groups (Nicol, 1978a). The results of his analysis showed clearly that habitat has a far greater effect than do phylogenetic relationships on the water turnover rate of a mammal as measured in the laboratory. This is in contrast to his conclusion about BMR (Nicol, 1978a). Thus, despite the doubts that have been expressed about the ecological significance of water turnover values obtained under *ad libitum* conditions (e.g. Denny & Dawson, 1975b; Nagy *et al.*, 1976), when taken over a wide range of species it does seem as if standard water turnover rates can be useful in separating 'desert hardy' species from others, as suggested by Macfarlane (1965).

Part of the controversy over Macfarlane's (1965) contention probably stems from the fact that the present habitats of some species may be quite different from those in which the species evolved. As Macfarlane (1976) said:

> Evolution in wet areas is associated with high turnover rates and low salt tolerance, while desert derivation goes with low rate functions and high salt tolerance. This basic ecophysiology changes slowly, and animals that migrate to different environments may retain ancient patterns in areas where they seem inappropriate – so that cattle keep their high rates of energy and water use in arid zones, or llamas remain low in energy and water turnover after three million years in cool or wet environments.

This may also help to explain the very high water turnover rates found in *Sminthopsis crassicaudata* by Kennedy & Macfarlane (1971) and Morton (1980a). This species is found in both xeric and mesic habitats in southern Australia, and yet seems inappropriate in the former.

Conclusion

It seems that, despite some influence of climate on the basal metabolic rate of some marsupials, and of some eutherians, the average BMR of marsupials is below that of the Eutherians.

The consequences of a low BMR were examined in relation to requirements of marsupials for energy, protein and water. In most instances there appears to be a close relationship between maintenance requirements for these nutrients and basal metabolism, although habitat has a far greater effect on the standard water turnover rate of an animal than on its BMR. Other nutrients such as vitamins and minerals will be examined in a later chapter. In the next six chapters we will see how these nutrient requirements are satisfied in various families of marsupials. The families will be grouped according to their food habits and mode of digestion, beginning with the carnivores.

2

Carnivorous marsupials

Although it is convenient to divide animals for the purpose of describing their dietary habits into one of three broad groups, viz. carnivores, omnivores and herbivores, it is often difficult to define the limits to each category. Thus many carnivores are seen to eat some plant material, either regularly or seasonally when prey species may be scarce or unavailable. For instance, although *Antechinus stuartii* is regarded as a carnivore, Fletcher (1977) found that in a sclerophyll forest in her study area in north-eastern New South Wales, plant material, mainly flowers of epacrid species, constituted up to 41% of the diet of *A. stuartii* during winter when insects, their principal food item, were present in relatively small numbers. Conversely, the tiny Honey Possum (*Tarsipes spencerae*) has a highly specialised, long, tube-like mouth and an extensible tongue brushed at the tip that it uses to collect nectar and pollen, its main food items; however, it also takes small, soft insects, at least in captivity (Vose, 1973).

Thus it is not always easy to defend marginal examples of carnivory or herbivory. Nor is it always possible to state whether a particular species is a strict carnivore or not. In the case of marsupial carnivores this is often at least partly due to a lack of sufficiently detailed information on the field ecology of the animal. For the purposes of this discussion a carnivorous species is one which eats mainly, though not necessarily only, animal material. This animal material may consist of either vertebrate or invertebrate prey, or both. Thus insectivory is not distinguished from other forms of carnivory.

Carnivore diets are characterised by a high content of protein, water, vitamins and minerals, a variable amount of fat, and a low level of carbohydrate. The high ratio of protein to carbohydrate in the diet means that little hexose is absorbed from the gut, and instead the animal's glucose and energy requirements are met largely from amino acids. Thus, compared with omnivores, the activity of the gluconeogenic enzymes pyruvate kinase and phospho-enol-pyruvate carboxykinase in the liver of carnivores is always high (Rogers, Morris & Freedland, 1977). Transaminase activity is also consistently high and as a result the maintenance protein requirements of strict carnivores are usually higher than those of omnivores, at least among the Eutheria. In the absence of information to the contrary the same is assumed to hold among marsupials.

The other characteristic of carnivore diets is the high digestibility of muscle and viscera, but very low digestibility of the exoskeletons of insects and other invertebrates, the teeth and some bones of vertebrates and bird feathers and mammalian hair. The fibrous proteins of feathers and hair, principally keratins, are extremely resistant to proteolytic attack. This means that examination of scats (faeces) of carnivores is a good way of working out their dietary habits. The scale arrangement and other features of mammalian hair are excellent diagnostic tools, and keys for the identification of mammalian hair are available (e.g. Brunner & Coman, 1974). The task is made more difficult among insectivores by the habit of many small species of eating only the softer parts of insects. For example planigales (*Planigale* spp.) dextrously use their forepaws to manipulate the hard exoskeleton of the prey so that it can be discarded.

The marsupial carnivores belong to one of two groups. The first group consists of two American families, the Caenolestidae (the shrew-opossums, seven species) and some members of the Didelphidae (the opossums, 70 species); other didelphids such as *Didelphis marsupialis* and *D. virginiana* are omnivorous rather than carnivorous. Hunsaker (1977) described *D. virginiana* as 'probably one of the most omnivorous vertebrates known. Food consumption apparently is based on whatever food is most readily available.'

The second group of carnivores belong to one of the Australian families Dasyuridae, Myrmecobiidae, Notoryctidae and Thylacinidae. The three latter families are monospecific. Of these, the third contains the Thylacine or Tasmanian Tiger (*Thylacinus cynocephalus*) which is probably extinct (Archer, 1979). The other two contain the insectivorous Numbat (*Myrmecobius fasciatus*) and Marsupial Mole (*Notoryctes typhlops*) respectively. In contrast, the family Dasyuridae contains 49 species, ranging in size from the tiny shrew-sized planigales (5 to 10 g) to the 10 kg Tasmanian Devil (*Sarcophilus harrisii*).

Diet studies

The Caenolestidae are usually described as shrew-like insectivores. Osgood (1921) found mainly insect and arachnid remains in the stomachs of three *Caenolestes obscurus* that he examined, and Hunsaker (1977) listed caterpillars, beetles, ants, centipedes and also spiders as staple food items of the caenolestids.

However, Kirsch & Waller (1979) considered from their observations of the feeding behaviour of a captive male *C. obscurus* when given live rats that this caenolestid may also be predaceous on small vertebrates: 'The animal would move toward a rat, sniffing vigorously, seize and lift the rat with its forepaws or pin it to the substrate, and bite it several times quickly with its incisors. The caenolestid would then commence eating the rat by biting off a section of the head with its cheek teeth and take successive bites posteriorly.' The few caenolestid stomachs that Kirsch & Waller (1979) opened in the field were empty, and thus they were not able to confirm their observations on captive *C. obscurus* in the wild.

The Caenolestidae possess peculiar lip flaps (Osgood, 1921) which have been variously suggested to function to hold, convey inwards, or eject food items. Kirsch & Waller (1979) suggested from their observations on captive *C. obscurus* that they may help prevent the sensory vibrissae and fur at the side of the mouth from becoming clogged with blood and dirt, as might result from the caenolestid's method of feeding, and also prevent dirt from entering the mouth.

Among the Didelphidae, *Marmosa* spp. (the murine or mouse opossums) are primarily insectivorous, but will also eat fruit, snails, birds, eggs, lizards, other small animals and carrion. Captive animals held by Enders (1935) died when fed only fruit, but the addition of insects to the diet resulted in good survival rates. Although Hunsaker (1977) found that *Marmosa* will survive and grow on commercial dog or cat diets, for successful reproduction these diets had to be supplemented with fresh fruit, meat and insects. The larger *Chironectes minimus* (the Water Opossum) is almost entirely aquatic, has webbed feet, and is found only in forested areas that have streams and rivers where it hunts for food. It is predominantly carnivorous, its diet consisting of insects, crustaceans, fish and frogs, although it may also eat some aquatic vegetation and fruit (Hunsaker, 1977). *Philander opossum* (the Grey Four-eyed Opossum) is a 250–400 g carnivore, feeding on small mammals, birds, bird eggs, carrion, and some fruit. It is primarily terrestrial, but is often found near streams and swamps (Hunsaker, 1977).

The diets of the Thylacine and the Marsupial Mole have been little studied. Troughton (1965) related the observation of A. G. Bolam, a keen naturalist, that Marsupial Moles in their desert habitat in central Australia lived on worms and insects, although it is possible that yam-like roots are also eaten. The natural food of the Thylacine consisted of wallabies and other smaller marsupials, rats, birds, and possibly lizards and Echidnas (Troughton, 1965). Its alleged attacks on domestic sheep and poultry after the advent of European settlement resulted in what appears to have been a successful extermination of the species.

The ecology and feeding habits of the Numbat have been reported by Calaby (1960). Its preferred habitat in south-west Western Australia is eucalypt woodland (chiefly *Eucalyptus redunca*) in which the heartwood of most trees has been eaten out by termites and the woodland floor is strewn with fallen hollow limbs and logs.

The Numbat is diurnal and generally solitary. From scat analysis Calaby (1960) determined that its diet consisted mainly of termites scratched from the upper 5 cm of soil. All species of termites were eaten, roughly in proportion to their abundance and availability. Ants made up to 15% of the diet, but the bulk of the ants eaten were the small predatory species which are probably ingested incidentally when they swarm in to prey on the exposed termites; Numbats apparently did not deliberately search for ants (Calaby, 1960). Other invertebrates were almost completely absent from Numbat scats.

Among the Dasyuridae the smaller members such as *Planigale* spp. and *Antechinus* spp. are almost wholly insectivorous, while the larger members take vertebrate prey as well. The most detailed dietary studies on the small Dasyuridae have been those of Fletcher (1977) with *Antechinus stuartii* and Morton (1980c) with *Sminthopsis crassicaudata*. Morton's (1980c) study compared the ecology and diet of *S. crassicaudata* at two extremes of its range, at Fowler's Gap in western New South Wales in the arid zone, and at Werribee in Victoria, a mesic environment.

Analysis of the stomach contents of *S. crassicaudata* for prey items was found to be impracticable because the contents were so highly triturated. Study of diet was therefore restricted to scat analysis. Morton (1980c) was careful to note that interpretation of his scat analysis data was subject to bias from three sources: (a) some invertebrate remains were easily recognisable because of their structural strength, e.g. arachnid leg parts, while others were easily overlooked; (b) not all possible prey items have structures which would appear in the faeces, e.g. earthworms (Lumbricidae) were found in stomach contents but were not recorded in faecal material;

and (c) some prey were probably dismembered before being eaten, and hard legs, wings and mouthparts may be discarded and not appear in the faeces, as has already been mentioned for planigales (page 28).

Given these qualifications, it is still possible to make some conclusions about the diet of *S. crassicaudata* in the two field study areas. Although the number of species identified in scats was greater in the arid zone (80 versus 45 at Werribee), the diets were largely similar. Spiders appeared to be an important year-round food source, as were grasshoppers (Acrididae), crickets (Gryllidae) and weevils (Curculionidae). Cockroaches (Blattodea) and slaters (Isopoda) were absent, and ants (Formicidae) were found infrequently and most often in winter. Vertebrates were rarely eaten. The vertebrate bones that were found were very finely splintered and unidentifiable. In view of the absence of hair in the scats, Morton (1980c) thought that they were most likely from frogs or reptiles rather than mammals. Some seeds and other plant material were sometimes present, but seeds were usually undigested. *S. crassicaudata* is thus almost completely insectivorous (Morton, 1980c).

In her field study of *Antechinus stuartii* in north-eastern New South Wales, Fletcher (1977) analysed a total of 520 scat samples collected from live traps, and compared the percentage occurrence of arthropods in scats with their percentage occurrence in pitfall traps throughout the year. Representatives of the Coleoptera (beetles), Araneae (spiders) and Blattodea (cockroaches) were the major constituents of the diet of *A. stuartii*. Other groups identified were Hymenoptera (ants), Orthoptera (grasshoppers and crickets), Diptera (flies), Myriapoda (centipedes and millipedes) and Dermaptera (earwigs).

With the exception of carabid beetles and non-anthropods, the components of the diet of *A. stuartii* followed closely the group composition of the pitfall captures. This led Fletcher (1977) to describe *A. stuartii* as an opportunistic feeder and that the active ground-dwelling arthropods were most important to its diet. She compared the feeding behaviour of *A. stuartii* with that of shrews (*Sorex* spp.) which will eat practically any small animal they encounter; dietary differences within shrew species reflect differences in availability which change with season, weather and habitat (Pernetta, 1976).

The most striking difference between habitats in Fletcher's (1977) study was in the taking of plant material by *A. stuartii*. Some plant material was eaten occasionally throughout the year in all habitats, perhaps unintentionally. However, all animals sampled in wet sclerophyll forest (tall forest) in late winter (when insect availability was minimal) were eating *Acrotriche aggregata* flowers. The eating of these flowers in other habitats followed the pattern of this shrub's distribution. When the flowers were taken they had obviously been picked from the branches since they were being eaten before any had fallen off the shrubs.

The diet of *Antechinus swainsonii* was found by Hall (1980) to be quite similar to that of sympatric *A. stuartii* in southern Victoria; the only differences were the more frequent occurrence of weevils (Curculionidae) in the faeces and stomach of *A. stuartii*, and the greater size range of prey taken by the larger *A. swainsonii*. Hall (1980) concluded that both *Antechinus* species were generalists and opportunistic feeders.

The Tasmanian Devil is the best studied of the larger Dasyurids. Because of its poor reputation among farmers for killing domestic stock Guiler (1970) made a careful study of the food preferences of three different devil populations on the basis of both qualitative analysis of scats and stomach contents and quantitative analysis of stomach contents (Table 2.1).

The population at Corinna depended almost exclusively on the wallabies *Macropus rufogriseus* (Red-necked Wallaby) and *Thylogale billardieri* (Red-bellied Pademelon). The other two populations made use of a wider food spectrum, including introduced species such as sheep and rabbits. Avian material was also present in the stomach contents, possibly the result of nocturnal predation. Tasmanian Devils, especially young animals, can climb trees and the bird remains were found mainly in the stomach of younger animals.

Thus it is clear that devils will utilise introduced as well as native species for food, depending on availability. This does not necessarily mean that the animals were killed by devils, since the presence of blowfly larvae (maggots) in some samples indicates that they were eaten as carrion (Green, 1967). Indeed, Buchmann & Guiler (1977) remarked on 'a general ineptitude among devils for killing'. Injured or moribund sheep may well be taken, and there is no doubt that devils are able to take quick and opportunistic advantage of natural mortality, and of cast lambs and offal. Thus they appear to be versatile scavengers rather than hunters.

The diets of the Eastern Quoll (*Dasyurus viverrinus*) and the Western Quoll (*D. geoffroii*) (Plate 2.1) are much more varied than that of *Sarcophilus harrisii*. Troughton (1965) was of the opinion that *D. viverrinus* fed mainly on insects, lizards and small birds and mammals including mice, rats and young rabbits. The importance of insects in the diet of *D. viverrinus* was confirmed by Blackhall (1980) in southern Tasmania, but plant material, mainly grasses and herbs, consistently occupied the greatest proportion of the faecal samples examined (28–41%). Other components

Table 2.1. *Principal food components in the diet of* Sarcophilus harrisii *at three sites in Tasmania; values given as percentages*

Site	No. of samples	Macropus rufogriseus	Thylogale billardieri	Vombatus ursinus	Pseudocheirus peregrinus	Birds	Sheep	Rabbits
Corinna	26	60.9	26.9	0	0	0	0	0
Glen Huon	17	17.6	0	0	17.6	5.8	0	41.1
Cape Portland	55	27.4	0	25.4	0	16.0	25.0	0

After Guiler (1970).

included fruit, mainly blackberries (*Rubus fruticosus*), and hair from house mice and Eastern Quolls. Blackhall (1980) thought that the Eastern Quoll hair probably originated mainly from loose hair shed or rubbed off in the traps used for capturing the animals. Unfortunately no stomach samples were taken. Thus it is likely that the results are heavily biased toward indigestible material such as vegetable matter and insect exoskeletons. Nonetheless the conclusion that considerable quantities of plant material are eaten by *D. viverrinus* is probably valid. Little avian material was detected in scats, indicating that *D. viverrinus* feeds almost exclusively on the ground.

In contrast, the larger Tiger Quoll (*Dasyurus maculatus*) is well known for its climbing ability, and its diet includes birds and their eggs along with other items including reptiles and mammals such as rabbits and small wallabies (Troughton, 1965). Similarly the phascogales (*Phascogale* spp.) are very active arboreally. Their raids on domestic poultry make them unpopular with many farmers, but little is known of their dietary habits in the wild. Gould stated that in the stomachs of some were found remains of beetles and what appeared to be a species of fungus (Troughton, 1965).

Plate 2.1. *Dasyurus geoffroii*, the Western Quoll, one of the larger carnivorous marsupials. (Ray Williams.)

Digestive tract morphology

Despite problems of definition, carnivores are universally distinguished from nearly all omnivores and herbivores by their gut morphology. Because of the generally highly digestible nature of food consumed, the overall impression of the carnivore digestive system is one of simplicity. The stomach is usually simple, without diverticulae or development of a forestomach fermentation region. The small intestine is short, as is the large intestine. The colon is usually of similar diameter to that of the small intestine.

Osgood's (1921) monographic study of *Caenolestes obscurus* includes a detailed description of the alimentary canal. The stomach (Fig. 2.1) is roughly elliptical in shape, and can be divided into the oesophageal, cardiac and pyloric regions. The oesophageal region is the smallest and the mucosa is organised into thin lamellae surrounding the opening of the oesophagus and extending along the lesser curvature to a point almost midway between the oesophagus and pylorus. The cardiac region is characterised by a highly differentiated glandular area that forms a compound cardio-gastric gland, reminiscent of the cardio-gastric gland of the Wombat and Koala to be described in Chapter 4. It is well developed laterally, being separate towards the pylorus but uniting around the cardiac side of the oesophagus. From the exterior it is clearly evident through the thin muscular walls. From the interior it appears as a thick,

Fig. 2.1. Stomach of *Caenolestes obscurus*. After Osgood (1921).

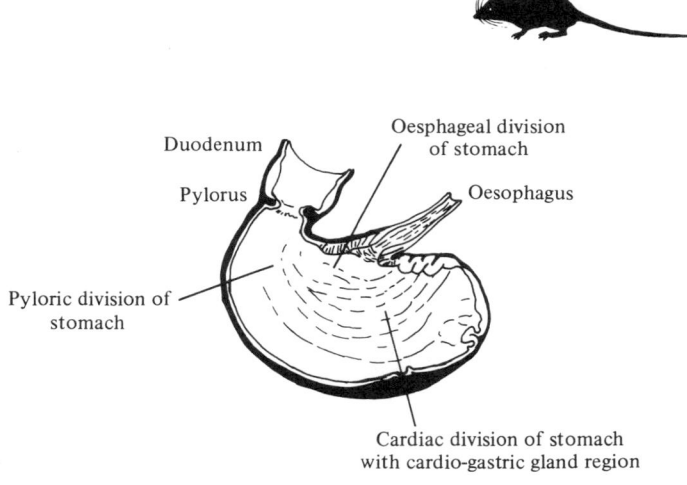

elevated, well circumscribed area of glandular tissue enclosing 40–60 slit-like openings. The pyloric region is quite smooth and undifferentiated and is continuous with the area along the greater curvature of the stomach.

The function of the cardio-gastric gland in *Caenolestes* is unknown, but its presence separates this species from all other carnivorous marsupials, both American and Australian.

The small intestine is relatively short, being 340 mm in an adult animal dissected by Osgood (1921). The duodenum appeared to have slightly thicker walls than the distal small intestine. The large intestine is very short, less than 10% of the length of the small intestine, and the division between colon and rectum is not easily seen. The caecum is very small, 'scarcely more than a vermiform appendix' (Osgood, 1921), and only 5 mm in length. Its mucosal lining resembles that of the ileum in being villous, in

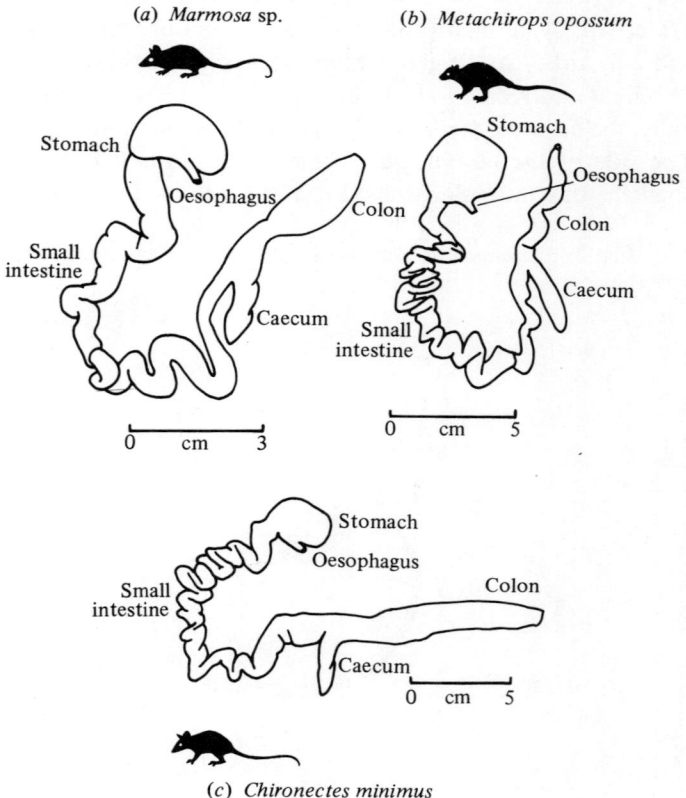

Fig. 2.2. Digestive tracts of *Marmosa* sp., *Metachirops opossum* and *Chironectes minimus*. A small caecum is present in each species. After Schultz (1976).

(a) *Marmosa* sp.

(b) *Metachirops opossum*

(c) *Chironectes minimus*

contrast to that of the colon which has a mucosal surface organised into large longitudinal folds.

Schultz (1976) has provided drawings of the digestive tracts of three other American carnivores, *Marmosa* spp., *Chironectes minimus* and *Metachirops opossum* (= *Philander opossum*). These are shown in Fig. 2.2.

The digestive tracts of these three didelphids have in common a simple stomach without any development of a specialised cardio-gastric gland, a short small intestine, an even shorter large intestine, and a small though distinct caecum. Barnes (1977) has described the anatomy of the digestive system of *Marmosa robinsoni*. The stomach is simple with glandular mucosa throughout, and there is minimal differentiation of cardiac and fundic areas. The pyloric region is mucus-secreting, and the pylorus is clearly defined externally by the presence of a white collar of duodenal or Brunner's glands.

These glands are peculiar to mammals, but within the Mammalia have been found in all of the Monotremata, Marsupialia and Eutheria (Krause, 1972). They secrete an alkaline fluid containing mucin which protects the proximal duodenal mucosa from the mechanical trauma of digesta moving through the intestinal tract and from the ulcerating effects of acid-pepsin secreted by the stomach. (The monotremes differ from the other mammalian superorders in that their stomach epithelium is non-glandular throughout (Griffiths, 1978) and Brunner's glands are confined to the submucosa of the distal stomach (Krause, 1970, 1971). Thus there is no peptic digestion in the monotreme stomach, nor is there a distinct pylorus, although there is an abrupt change from squamous epithelium to glandular mucosa at the proximal end of the duodenum.)

In *Marmosa robinsoni* the short, uncomplicated small intestine has two histologically definable regions, the duodenum and the ileum. The duodenum is characterised by a large calibre and long, finger-like villi without duodenal crypts. The ileum is of slightly smaller calibre but has a larger lumen owing to its shorter villi. Glandular crypts are short and goblet cells (mucin-secreting) gradually increase in number toward the ileocaecal valve. In contrast to *Caenolestes*, the short caecum of *Marmosa* is histologically similar to the colon in that villi are completely absent. Crypts lined with abundant goblet cells form the mucosa. The colon lacks haustra (non-permanent sacculations) or other special features.

Thus *Marmosa*, and presumably *Chironectes* and *Philander* also, differ from *Caenolestes* in both the absence in the stomach of any specialised cardio-gastric gland and the presence in the hindgut of a more than vestigial caecum. Osgood (1921) regarded the short colon of *Caenolestes* as a primitive condition, and the tiny caecum a secondary one.

38   2 Carnivorous marsupials

The digestive tracts of the Thylacine (Mitchell, 1916), the Marsupial Mole (Mitchell, 1916; Schultz, 1976), the Numbat (Schultz, 1976) and all of the Dasyuridae that have been examined have no caecum at all. These include *Sminthopsis* spp., *Antechinomys laniger* (Schultz, 1976), *Dasyurus maculatus* (the Tiger Quoll), *Phascogale tapoatafa* (the Brush-tailed Phascogale) and *Dasyuroides byrnei* (the Kowari). *Notoryctes*, *Myrmecobius* and the two dasyurid species examined by Schultz (1976) are shown in Fig. 2.3. *Dasyurus maculatus* and *Phascogale tapoatafa* are shown in Fig. 2.4.

Except for the absence of a caecum the digestive tracts shown in Figs. 2.3 and 2.4 are similar to those of the three didelphids just described, and to those of most eutherian carnivores. Owen (1868), one of the early English anatomists interested in comparative aspects, considered that the stomach of carnivorous marsupials was 'relatively much more capacious'

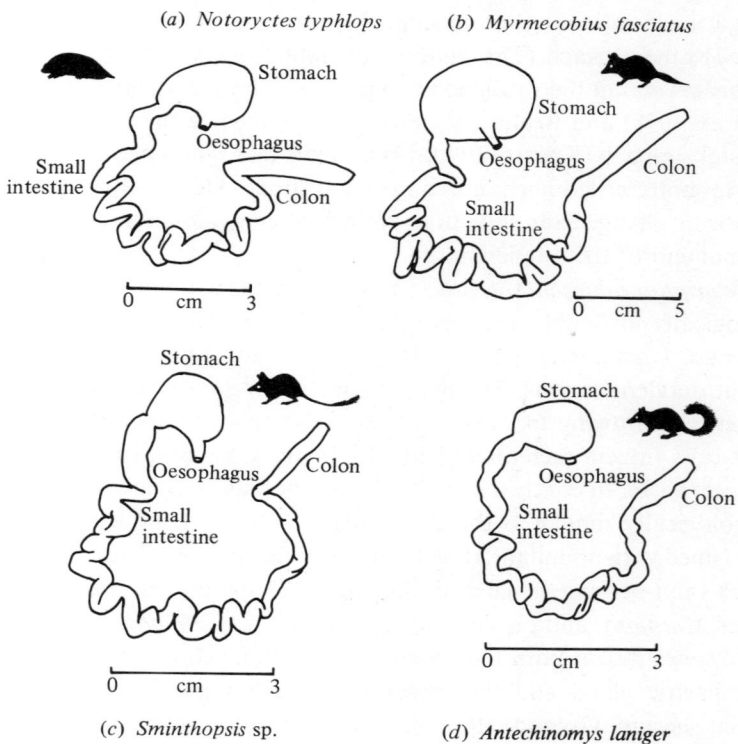

Fig. 2.3. Digestive tracts of *Notoryctes typhlops*, *Myrmecobius fasciatus*, *Sminthopsis* sp. and *Antechinomys laniger*. These species are distinctive in that they lack a caecum. After Schultz (1976).

(a) *Notoryctes typhlops*   (b) *Myrmecobius fasciatus*

(c) *Sminthopsis* sp.   (d) *Antechinomys laniger*

and 'better adapted for the retention of food' than in the Carnivora, but the detailed information necessary to test this statement is lacking.

The lack of a caecum in the Dasyuridae, Myrmecobiidae, Thylacinidae and Notoryctidae is considered to be an example of secondary loss of a primitive organ (Mitchell, 1905), in the same way that Osgood (1921) regarded the very small caecum of *Caenolestes* to be a secondary condition. Hume & Warner (1980) argued that the first mammals in the Jurassic, which are thought to have been all small, nocturnal insectivores, most probably had a small caecum, derived from reptilian ancestors, and homologous with the paired caeca of birds. Mitchell (1905, 1916) described what he considered to be 'a second caecum, much shorter than the first, but quite distinct' in two macropodine species (see Fig. 5.2), and referred to 'a paired condition of the caeca being a primitive mammalian character'. Secondary loss of this primitive organ appears to have occurred not only in some carnivorous marsupials but also in some eutherian carnivores such as the Mink (*Mustela vison*).

Digestive function

Because of their short, uncomplicated digestive tract the rate of passage of food residues through eutherian carnivores is rapid. Average passage times of 142 minutes have been recorded with Mink (Sibbald,

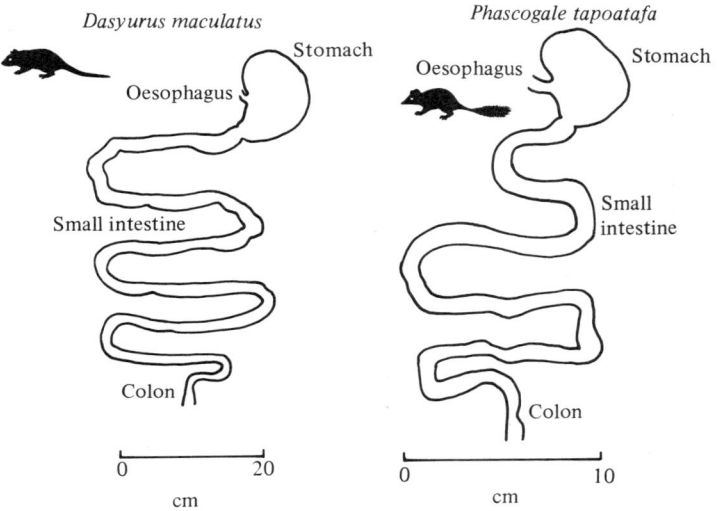

Fig. 2.4. Digestive tracts of *Dasyurus maculatus* and *Phascogale tapoatafa*. These species also lack a caecum, and the digestive tract is very short.

Sinclair, Evans & Smith, 1962), but in other species such as the Raccoon (Clemens & Stevens, 1979) 95% excretion times for both fluid and particulate markers were close to 24 hours. Cowan *et al.* (1974) examined rate of passage in the small (50 g) marsupial carnivore *Antechinus swainsonii* by feeding a 50–50 mixture by weight of ground house mice and termites (*Nasutitermes exitiosus*). The indigestible termite heads were counted in faeces collected either one- or two-hourly after feeding. Food offered at 1700 h was consumed over the next 6–8 hours, with intervals of rest intervening between feeding activity. The pattern of appearance of the termite heads is shown in Fig. 2.5.

The first termites appeared within 3 hours, 90% had been eliminated within 12 hours, and 99% within 18 hours. Thus, although slower than in the Mink, rate of passage of food residues through *A. swainsonii* is comparable to estimates from the eutherian Raccoon and dog examined by Clemens & Stevens (1979).

When fed exclusively on a diet of ground house mice, apparent dry matter digestibility in *A. swainsonii* averaged 80%, and apparent energy digestibility averaged 87%, illustrating the ease with which carnivore diets can be digested despite rapid passage rates. Green & Eberhard (1979) fed Tasmanian Devils (*Sarcophilus harrisii*) and Eastern Quolls (*Dasyurus viverrinus*) on whole dead laboratory rats and recorded apparent dry matter digestibilities of 79% and 81% in the two species respectively, and apparent energy digestibilities of 87% and 89%, remarkably similar

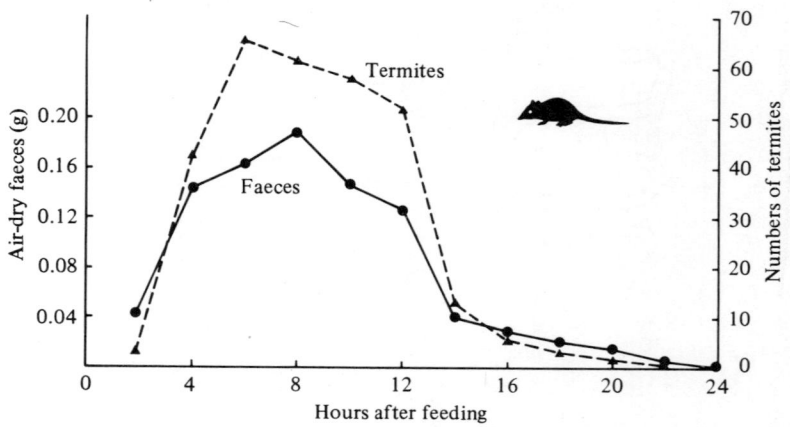

Fig. 2.5. The pattern of appearance of termite heads in the faeces of *Antechinus swainsonii*. The fast rate of passage through *A. swainsonii* is comparable with that in several eutherian carnivores. After Cowan *et al.* (1974).

to the estimates of Cowan *et al.* (1974) in *A. swainsonii*. Apparent digestibility of the dry matter and energy of *Tenebrio* larvae by *A. stuartii* was 84% and 87% respectively (Nagy *et al.*, 1978).

Specific adaptations in small dasyurid species

Small dasyurids are found in all the main habitats in Australia, including desert, despite the inability of many species to maintain a steady body temperature at ambient temperatures of 40 °C (Robinson & Morrison, 1957). Their adaptations for survival include avoiding direct exposure (all are nocturnal and most are fossorial), a marked diurnal temperature cycle, and torpor (Tyndale-Biscoe, 1973). Torpor (a state of reduced metabolism) is used by many small mammals to reduce their energy expenditure and thus their total demands for energy, water and other nutrients.

The influence of small body size on energy requirements is easily seen among the Dasyuridae, many of which weigh less than 50 g. Per unit of body weight, small animals consume oxygen faster than do large animals. Thus adult man eats the equivalent of about 1% of his body weight each day, but *Antechinus stuartii* (26 g body weight) in the field consumes 60% of its body weight in arthropods each day (Nagy *et al.*, 1978).

There is, in body size, a balance of advantages and disadvantages in energy relations. A small animal cannot survive more than a few days without food, so that in adverse times a small-sized species will decline rapidly as its members succumb to the lack of food and/or adverse environmental conditions. Members of larger species can withstand adversity for much longer. On the other hand, under favourable conditions the small-sized species convert food more rapidly, and the population can proliferate faster than the large species.

These energy considerations explain why small-sized mammals and birds are restricted in their dietary range to easily digested, concentrated foods such as insects and small vertebrates; nectar and seeds can also be included in this category. Many small mammals also reduce their total energy demand by entering torpor. Hudson & Bartholomew (1964) have proposed three degrees of reduced metabolism. *Facultative torpor* is the least developed form. There is a daily fluctuation in body temperature and basal metabolism, but not enough to stop feeding or excretion. At the low point in the diurnal cycle the animal need only stop shivering to cool to a body temperature characteristic of torpor. Such facultative torpor may occur in response to food shortage, but the animal cannot survive more than a few hours in torpor nor tolerate a body temperature less than about 15 °C.

The next stage is *obligate daily torpor*, as seen in desert-adapted species that avoid high daytime temperatures and consequent water loss by entering a burrow. This immediately reduces evaporative water loss. By lowering body temperature to that of the burrow a further reduction is achieved. Together, these factors can reduce total water loss to 3.5% of that on the surface at normal body temperature, a very significant factor in the animal's water economy (Tyndale-Biscoe, 1973).

The final stage is true *hibernation*, in which torpor lasts for extended periods. This is an adaptation to low ambient temperatures. Hibernating species of more than 200 g body weight can usually store sufficient fat to provide their total energy requirements during hibernation. However, small species are unable to carry sufficient reserves, and are obliged to eat and drink during regular periods of arousal. Arousal also provides an opportunity for animals to excrete potentially toxic metabolic products, especially urea. Urea accumulates during torpor, but arousal is necessary to excrete it, since the kidney cannot function at the low temperatures of torpor. Thus all hibernating animals arouse from torpor occasionally, and some regularly, even though arousal is energetically expensive, being equivalent to about ten days of torpor.

The desert-dwelling dasyurids have been studied primarily because of their energetic relationships, and only indirectly because of their carnivorous dietary habits. However, Haines *et al.* (1974) compared water turnover in Australian desert dasyurids and murid rodents. The comparison is interesting because the dasyurids, being essentially carnivorous, obtain appreciable amounts of water in their food, while the granivorous rodents obtain very little water from this source. On the other hand, the carnivorous dasyurids are likely to require more water in order to excrete the nitrogenous end products from their high-protein diet.

Water turnover ($ml \cdot kg^{-0.80} \cdot day^{-1}$) in two dasyurids, *Dasycercus cristicauda* and *Dasyuroides byrnei*, was similar to that in four desert rodent species (*Notomys alexis, N. cervinus, Pseudomys australis, P. desertor*). In contrast, the dasyurid *Sminthopsis crassicaudata* expended water at rates comparable with non-desert species. This concurs with Kennedy & Macfarlane's (1971) report on *S. crassicaudata*, and with Morton's (1980a) laboratory results.

When Haines *et al.* (1974) removed drinking water from *Dasyuroides* fed fresh meat (60% water) its water turnover fell by 40%, enabling it to remain in water balance. *Dasycercus cristicauda* also remained in water balance when fed fresh meat (68% water) without drinking water, but with no reduction in water turnover (Kennedy & Macfarlane, 1971); presumably *Dasycercus* obtained all its water from the diet (Schmidt-Nielsen &

Newsome, 1962). However, *Sminthopsis crassicaudata* could not survive this regime for more than two to four days (Kennedy & Macfarlane, 1971). Morton (1980a) found that *S. crassicaudata* could subsist without drinking water when fed a diet of insects, but only if the insects contained more than 60% water. Evidently *Dasyuroides* and *Dasycercus* can concentrate their urine sufficiently to achieve water balance despite the need to excrete large amounts of urea formed from their high-protein diet. Indeed, Schmidt-Nielsen & Newsome (1962) found that in *Dasycercus* urine the urea concentration reached 2610 mM, about eight times the plasma concentration. On the other hand, *S. crassicaudata* may require exogenous water in addition to that in fresh meat or insects to maintain circulation and renal excretion. Although Morton (1980a) showed that *S. crassicaudata* urine reached maximum total osmolarities approaching those of *D. cristicauda* when water was restricted, its concentrating ability with respect to urea is unknown. It is noteworthy that *Dasycercus* inhabits the most arid regions of central Australia, where free water is usually not available. *S. crassicaudata*, however, lives in habitats ranging from mesic grasslands to arid stony plains (Morton, 1980a) and appears to be adapted to intermittent water and food supply (Kennedy & Macfarlane, 1971).

Plate 2.2. *Sminthopsis macroura*, the Stripe-faced Dunnart, a small carnivore which readily enters torpor. (Ray Williams.)

When forced to minimise water intake the four desert rodents reduced water turnover by 70% (Haines *et al.*, 1974), reflecting their need to conserve water since they obtain little from their diet of dry seeds (approximately 12% water or less).

Torpor is probably used by *D. cristicauda* (Morrison, 1965), and by *S. macroura* (= *S. larapinta*) (Godfrey, 1968) to reduce its water loss in its burrow during daylight hours. *S. crassicaudata*, however, has been shown to maintain a much more constant body temperature during the day (Godfrey, 1968) and only enters diurnal torpor during winter (Morton, 1980c). This is presumably because it feeds almost entirely on small insects and other arthropods (Morton, 1980c), and therefore must spend more time at night on the surface hunting for food than must *D. cristicauda* and *S. macroura* (shown in Plate 2.2) which eat small vertebrates as well as insects (Godfrey, 1968; Tyndale-Biscoe, 1973). Crowcroft & Godfrey (1968) found in the laboratory that *S. crassicaudata* never became torpid if food was plentiful, even at ambient temperatures as low as 5 °C. However, in the absence of food the animals entered torpor, thereby conserving energy. When starvation was continued for more than two days the period of torpor lengthened and some animals failed to arouse and died. This suggests that, in contrast to *Dasycercus*, *S. crassicaudata* only enters torpor as an emergency measure, and minimises the effects of temporary food shortages by drawing upon the fat stored in its tail (Morton, 1980b) and by other energy-conserving strategies (see page 45).

Dawson & Wolfers (1978) were particularly interested in the metabolism and homeothermic characteristics of the smallest dasyurids, the planigales. Although the body weights of the three species studied (*Planigale ingrami*, *P. tenuirostris* and *P. gilesi*) were only 5.8–9.4 g, BMR was at the level predicted with the metabolic rate–body weight relationship for marsupials (see Chapter 1). This is in contrast to their eutherian counterparts, the shrews, which have metabolic rates much higher than would be expected from Kleiber's (1932) equation.

Body temperatures of resting planigales in thermoneutrality (33 °C) were 34.5 °C for the first two species, and 36.9 °C for *P. gilesi*. However, at lower temperatures metabolic rate increased sharply. At 14 °C metabolism was 10 times BMR, and even at 25 °C it was four times BMR. Thus the maintenance of body temperature obviously requires considerable metabolic activity. Planigales reduce this metabolic load and its consequent high food requirement by lowering body temperature when ambient temperatures fall. At 14 °C the body temperature of active animals fell to 32.0 °C. This lower body temperature would tend to reduce the gradient for heat loss and thus reduce the metabolic requirement for homeothermy.

In order to reduce this metabolic requirement still further, *P. ingrami*, the more tropical species, often entered torpor at ambient temperatures as high as 22–26 °C, even when food was available in excess; metabolic rate during torpor fell to 30–50% of BMR. *P gilesi*, on the other hand, a more temperate species, did not exhibit a pattern of torpor, even at ambient temperatures of 14 °C, unless food intake was restricted. In this behaviour it was similar to *S. crassicaudata* (Crowcroft & Godfrey, 1968).

Results from Morton's (1980a, b, c) detailed ecological study of *S. crassicaudata* suggest that three different energy-conserving mechanisms are used by this species as buffers against unpredictable periods of food shortage in the open environments in which it lives. These are torpor, nest-sharing and caudal fat storage. Torpor and nest-sharing (huddling) were observed in field animals only during winter, the season of lowest food abundance. Caudal fat stores reached a peak in autumn. Morton (1980c) interpreted this as an expression of the immediate energy balance of the animals; in spring and summer breeding animals store little fat because of the energetic demands of reproduction despite the fact that insect populations normally peak in these seasons. The significance of caudal fat storage thus appears to be limited. The amount of energy stored as fat in the tail (4–17 kJ) is less than that contained in the body (approximately 29 kJ) and small relative to the daily energy requirements of animals in the laboratory (71 kJ); it must be an even smaller fraction of the total energy requirements of free-living animals. Nevertheless, the existence of caudal fat storage in many small mammals shows that there are strong selective forces favouring energy storage, and the selective pressure seems to be short-term variability in insect abundance (Morton, 1980d). Thus, although only a small energy reserve, caudal fat storage, along with torpor and nest-sharing must be important in providing a buffer against unpredictable periods of food storage, especially in winter, the season of lowest insect abundance (Morton, 1980c).

The fact that all three energy-conserving devices were found in *S. crassicaudata* at both Fowler's Gap in the arid zone and Werribee in southern Victoria, together with the apparent lack of any specific adaptation in the energy and water metabolism of *S. crassicaudata* to aridity (Kennedy & Macfarlane, 1971; Morton, 1980a), suggested to Morton (1980c) that the biology of this small dasyurid includes a series of adaptations countering variability in food supply, and that the variability is a major problem facing a nocturnal insectivore inhabiting open environments, regardless of the aridity of the habitat.

Two other characteristics of *S. crassicaudata*, that is high mobility (the home range is greater than 600 m in diameter (Morton, 1980b), compared with 100–200 m for *Antechinus stuartii* in a rainforest habitat (Wood, 1970)), and repeated breeding within an uncertain optimum period (Morton, 1980c), are also consistent with habitats which are open and relatively unprotected from sudden changes in the weather and therefore from unpredictable short-term shortages in insect availability.

At the other end of the scale is *Antechinus stuartii*, which lives in well-watered forest habitats where insect availability also varies seasonally but in a highly predictable fashion (Fletcher, 1977). Related to this is a very unusual life history, with breeding occurring only once per year, in late winter. Wood (1970) followed an entire population of *A. stuartii* in a 0.8 hectare area of Queensland rainforest for three years. By repeated recapture of marked animals he was able to monitor the total numbers of *A. stuartii* and the structure of the population. As can be seen from Fig. 2.6 males completely disappeared from the population in spring, and reappeared in summer. In contrast, females all survived into their second and some into their third year. Males caught just prior to their disappearance showed a marked loss of weight compared with earlier captures, and Wood (1970) concluded that the disappearance was due to widespread death of males in spring, rather than to migration out of the area. This field study confirmed Woolley's (1966) findings from observations on a captive colony of *A. stuartii* at Canberra. She found a very precise synchrony in the onset of sexual maturity in young males and females. The females, both one- and two-year-old animals, come into oestrus in July and August. Oestrus lasts for ten days and coincides with the peak of sexual development in the males.

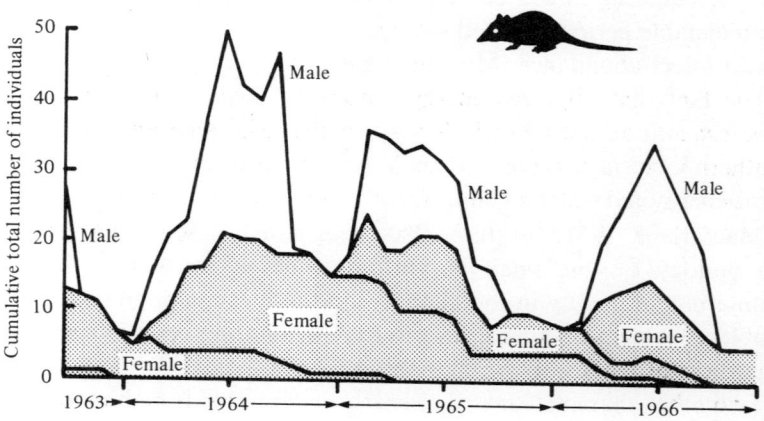

Fig. 2.6. Population changes in *Antechinus stuartii* in a Queensland rainforest. Note the complete disappearance of males in October each year. After Wood (1970).

By the end of August all females are pregnant, and the males have passed into a general decline, with loss of fur and body weight. In captivity they can be kept alive, but the degeneration of the testes is irreversible so they are quite sterile.

Woollard (1971) followed changes in body weight, energy intake and nitrogen balance of five male and five female *A. stuartii* over the period March–October. The animals were fed minced meat, and urine and faeces

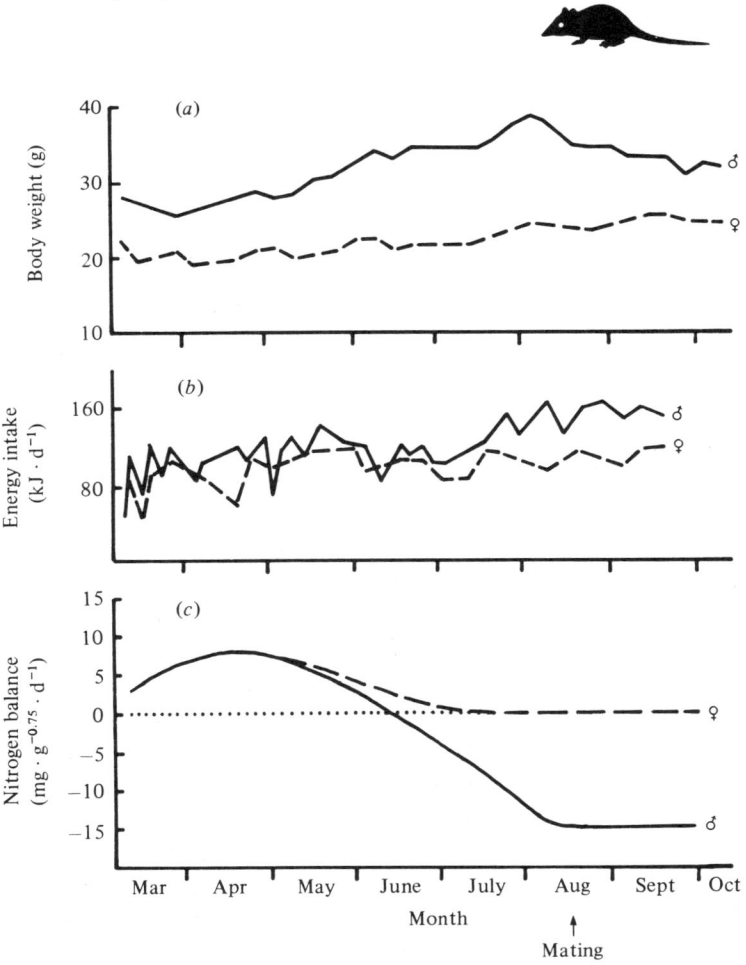

Fig. 2.7. Changes in body weight, energy intake and nitrogen balance of a laboratory colony of *Antechinus stuartii* during winter and spring. Note the dramatic effect of mating activity in males. After Woollard (1971).

were collected on blank newsprint on the cage floor. Woollard's (1971) results are summarised in Fig. 2.7. The pattern of body weight change was similar to that observed by Woolley (1966) in the laboratory and by Wood (1970) in the field. In young animals energy intake was slightly higher in males than in females, which can be explained by the higher body weights of the males. From June, energy intake of females stabilised while that of males increased. Nitrogen balance was positive and increasing in March and April, but thereafter declined. Females were close to balance from June, but, despite increased food consumption, males went into negative nitrogen balance in June, and remained in substantial negative balance until the end of the study in late September.

Subsequent work by Bradley, McDonald & Lee (1975, 1976) has elucidated the endocrine changes involved in the annual die-off of male *A. stuartii*. Male mortality appears to result from mating behaviour which induces large increases in plasma free corticosteroids with subsequent increases in skin hydroxyproline concentration (Barnett, 1974), disruption of the immune system, anaemia, haemoglobinuria, focal hepatic necrosis, gastrointestinal haemorrhage due to gastric and duodenal ulcers (Barker, Beveridge, Bradley & Lee, 1978) and other consequences (Lee, Bradley & Braithwaite, 1977).

Although the haematological changes appear moderate when compared with anaemias tolerated by larger animals, an associated increase in resting metabolism (Cheal, Lee & Barnett, 1976) is indicative of an increase in the cost of oxygen transport in the presence of the anaemia. Even moderate decreases in the oxygen-carrying capacity of the blood are likely to be taxing for small mammals with their relatively high oxygen demands.

At least five other *Antechinus* species – *A. swainsonii*, *A. flavipes*, *A. minimus*, *A. bellus* and *A. bilarni* – also appear to be monoestrous and show a post-mating mortality of males (Woolley, 1966), although the male mortality is most abrupt and is best described in *A. stuartii* (Lee *et al.*, 1977). All species breed in winter or early spring, and the litters are weaned in early summer between November and January. Increasing food availability in spring thus coincides with lactation (Lee *et al.*, 1977) and maximum food availability with the period of maximum food requirement, the few weeks after weaning (Fletcher, 1977). If a second litter followed immediately, weaning would occur in autumn and coincide with declining insect availability. These circumstances favour the production of only one litter in the spring of each year. Since the probability of survival from year to year is low in an animal of this size, selection may be expected to favour individuals which maximise reproductive effort during their first reproduction. It is this reproductive effort by males, including searching for females, prolonged and repeated copulation, and agonistic encounters with other males, which appears to trigger the endocrine changes described by Bradley *et al.* (1975, 1976), leading to their mortality (Lee *et al.*, 1977).

There is thus a fundamental difference between the life histories of the two small dasyurids *A. stuartii* and *S. crassicaudata* which can be related directly to differences between their respective habitats, principally in the variability in availability of insects, their staple food source.

Summary and conclusions

The marsupial carnivores belong to one of two groups, the American families Caenolestidae (the shrew-opossums) and some Didelphidae (the opossums), and the Australian families Dasyuridae, Myrmecobiidae, Notoryctidae and Thylacinidae. Good descriptions of the digestive tract have been published only for two American species, *Caenolestes obscurus* and *Marmosa robinsoni*. However, the digestive tract of all marsupial carnivores, in common with that of eutherian carnivores, appears to be relatively short and simple. As a consequence the rate of passage of food residues through the gut is rapid. Nevertheless, digestibility of the dry matter and gross energy of their natural food items by marsupial carnivores is high.

Some specific adaptations of small dasyurid species are described. These include the use of torpor, nest-sharing and fat storage to minimise the effects of temporary food shortage, and the ability of some desert species to concentrate their urine many-fold with respect to urea, the principal metabolic end product of their high-protein diets, in order to conserve water.

The remarkable life history of *Antechinus stuartii* is discussed in relation to the regular seasonal changes in availability of its food supply, and is compared with that of other dasyurids which are faced with a much less predictable food supply.

3

Bandicoots and other marsupial omnivores

Omnivory, by definition, includes ingestion of plant as well as animal material. This means that greater amounts of indigestible residues will be consumed by omnivores than by carnivores. This has several important nutritional consequences. One of these is the requirement for greater lubrication to protect the gut lining from physical trauma during digesta passage (Hume & Warner, 1980). Another is that plant residues provide an additional substrate for bacteria resident in the hindgut, primarily the caecum. Thus, compared with carnivores, we usually see in the omnivore digestive tract an increased caecal capacity, together with an increase in small intestinal length and in colon length, diameter or both.

Marsupial omnivores fall into three groups. The first consists of the families Peramelidae (bandicoots) and Thylacomyidae (bilbies). The second consists of some members of the New World family Didelphidae, of which *Didelphis* is the best known member. The third group consists of several Australian arboreal species which feed on a mixture of non-foliage plant materials and arthropods. These include *Dactylopsila trivirgata* (the Striped Possum), *Gymnobelideus leadbeateri* (Leadbeater's Possum), *Petaurus australis* (the Yellow-bellied Glider) and *P. breviceps* (the Sugar Glider). The smaller *Cercartetus nanus* (the Eastern Pygmy-possum) and *Burramys parvus* (the Mountain Pygmy-possum) can also be included in this group.

Bandicoots and bilbies

Bandicoots and bilbies are all terrestrial and mainly nocturnal small to medium sized marsupials found in Australia, New Guinea and nearby islands. Feeding habits appear to vary between locations, but all species can be classed as omnivorous. Heinsohn (1966) made a detailed study of two bandicoot species in Tasmania, *Perameles gunni* (the Eastern Barred Bandicoot) and *Isoodon obesulus* (the Southern Brown Bandicoot). The major food items taken by the two species were largely similar, and consisted mainly of earthworms, adult beetles, moth larvae and pupae, and scarab larvae. The only plant products extensively used by *Perameles* were ripe blackberries and boxthorn berries; stomachs of two *Perameles* collected in February were distended with berries. The only plant material taken by *Isoodon* in large quantities was boxthorn berries, which were heavily utilised when ripe in the late summer.

Wood Jones (1924) considered that *I. obesulus* in South Australia scratched at the roots of vegetation with the object of obtaining insects on the roots rather than the roots themselves. *Perameles eremiana* (the Desert Bandicoot) was similarly described as mainly insectivorous by Wood Jones (1924), but the diet of the now extinct Pig-footed Bandicoot (*Chaeropus castanotis* = *C. ecaudatus*) as omnivorous since it included grass, bulbous roots and grasshoppers. Ride & Tyndale-Biscoe (1962) described the habitat of *Perameles bougainville* (the Western Barred Bandicoot) on Bernier and Dorre Islands in Shark Bay, Western Australia, but did not include any comment on diet preference.

Although earlier workers had described the diet of the bilbies to consist of grass, bulbous roots, fruits and insects, Wood Jones (1924) doubted very much if any of the species was at all given to eating roots, grass or fruits:

> It is true that in districts where they live it is common to see little holes scratched around the roots of vegetation, but it is very doubtful if these are made in order to obtain roots. It seems much more likely that insects are the object of the search. In captivity I have been unable to persuade them to eat roots or fruit; but bread or cake, meat (raw or cooked), insects, snails, birds, and mice are all readily eaten.

On the other hand results presented by Johnson (1980) on the basis of microscopic analysis of scats collected from nine colonies of *Macrotis lagotis* (the Greater Bilby) in the desert region north-west of Alice Springs, Northern Territory, support the observations of the earlier workers. The food material consisted of plants (seeds, fruit, bulbs), insects (mainly ants, termites and beetles) and a small amount of fungus. There was no apparent preference for any particular food item, and Johnson (1980) concluded that the diet appeared to reflect the seasonal and temporary availability of the various components. Insects averaged 32% of the identifiable fragments in the faeces. In two colonies living near salt lakes bulbs of the sedge

52   3 Bandicoots and other marsupial omnivores

Cyperus bulbosus formed 61% and 57% of food material. The annual grass *Panicum australiense* formed 67% of identifiable fragments in one colony, and seeds of the desert trigger plant *Stylidium desertorum* formed 33% in another.

In captivity bilbies collected scattered seed, one grain at a time, with rapid movements of the tongue. Seed was temporarily stored in cheek pouches before mastication. Live house mice were ignored.

Thus there can be little doubt about the omnivorous dietary habits of both bilbies and bandicoots. The wide food spectrum of this group is reflected in their digestive tract anatomy and function. The gut of *Perameles nasuta* (the Long-nosed Bandicoot) is shown in Fig. 3.1, and that of *Macrotis lagotis* in Fig. 3.2.

In both species the stomach is simple, as is the small intestine. The most obvious feature of the hindgut is a caecum of moderate size. Schultz (1976) presented a specimen of *Perameles* sp. with a smaller caecum, but a colon of greater diameter (Fig. 3.3).

Parsons (1903) examined a specimen of *Chaeropus ecaudatus* and

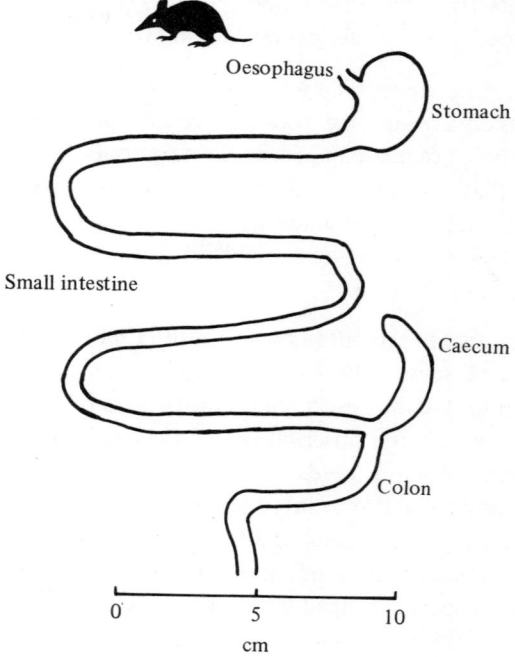

Fig. 3.1. Digestive tract of *Perameles nasuta*. The caecum is larger than in the carnivorous marsupials. After Harrop & Hume (1980).

Bandicoots and bilbies

Fig. 3.2. Digestive tract of the rare *Macrotis lagotis*. Drawn from a preserved specimen provided by K. A. Johnson.

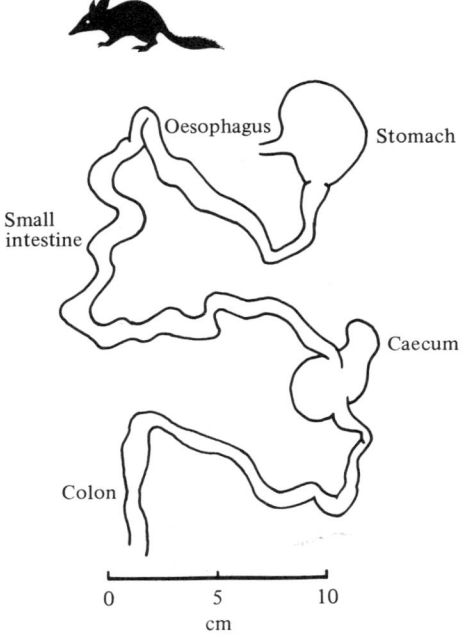

Fig. 3.3. Digestive tract of *Perameles* sp. After Schultz (1976).

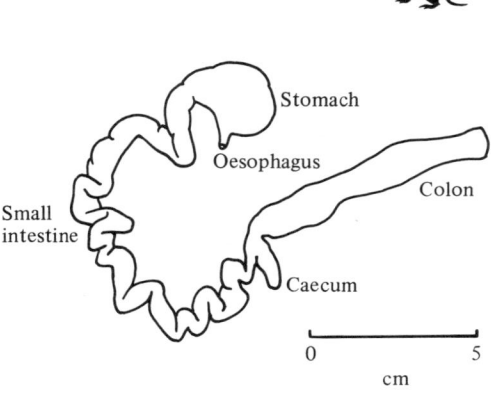

remarked on the great size of the cardiac portion of the stomach as well as 'the attempt at a marking off of a small secondary chamber or pyloric antrum'. The small intestine was 180% of body length, and the large intestine equal to the length of the body. The caecum was relatively longer than those shown in Figs. 3.1 and 3.3 for *Perameles*, suggesting that plant material probably constituted a significant part of this animal's diet. This accords with Wood Jones's (1924) description of the food habits of *C. ecaudatus* as being clearly omnivorous.

Since only one specimen of *Chaeropus* has been collected this century no studies of digestive function in the Pig-footed Bandicoot have been carried out. Kerry (1969), however, has measured the activities of several disaccharidases in the mucosa of the small intestine of *P. nasuta* and *I. obesulus* (Table 3.1). The activity of trehalase and cellobiase was substantial in both species. Trehalose is a disaccharide found only in insects, while cellobiose is formed during cellulose breakdown. Thus the bandicoots are adapted to utilising sugars of both insect and plant origin. By way of comparison *Antechinus stuartii* and *Dasyurus maculatus* both exhibited high levels of trehalase, but not of cellobiase, reflecting their predominantly carnivorous/insectivorous dietary dabits.

The rate of passage of digesta through the bandicoot gut might be expected to be a little slower than in carnivores, given their slightly longer digestive tract. Griffiths (see Waring, Moir & Tyndale-Biscoe, 1966) used different species of termites as passage markers in *Isoodon macrourus* (the

Plate 3.1. *Isoodon macrourus*, the Northern Brown Bandicoot or Short-nosed Bandicoot, an omnivorous marsupial. (Gordon Lyne.)

Table 3.1. *Intestinal disaccharidase activity in marsupials*

	Disaccharidase activity (units[a] per g wet weight of mucosa)					
	Maltase	Isomaltase	Sucrase	Lactase	Trehalase	Cellobiase
Family Dasyuridae						
Antechinus stuartii	43	—	10	0.1	8	0.1
Dasyurus maculatus	69	38	5	1.3	24	0.1
Family Peramelidae						
Perameles nasuta	19	12	5	0.7	11	0.3
Isoodon obeselus	—	—	0	—	4	0.2
Family Phalangeridae						
Trichosurus vulpecula	41	23	7	0.5	7	0.2
Family Petauridae						
Pseudocheirus peregrinus[b]	0.2	0.1	0.1	11.2	—	0.1
Family Phascolarctidae						
Phascolarctos cinereus	12	5	2	0.7	0	1.4
Family Macropodidae						
Macropus giganteus	0.3	0.1	0	0.1	0.1	0
Macropus giganteus[b]	0.6	0.1	0	5.6	—	0.1

After Kerry (1969).
[a] One unit of disaccharidase activity hydrolyses 1 μmol substrate$\cdot$min^{-1}.
[b] Pouch young.

Northern Brown Bandicoot) (shown in Plate 3.1) by examining collected faeces for undigested exoskeletons. Marker termites first appeared in the faeces 7 hours after ingestion, reached a peak concentration at 9 hours, and were completely eliminated in 29 hours. This indeed indicates a slightly longer total passage time in the omnivorous bandicoots than in carnivores such as eutherian mink (Sibbald et al., 1962) and the marsupial *Antechinus swainsonii* (Cowan et al., 1974).

The Opossums

The most successful didelphid is the Virginia Opossum, *Didelphis virginiana*. Its present range extends from southern Ontario, Canada to as far south as Nicaragua (Hunsaker, 1977). Within this vast area *Didelphis* is mainly confined to the more temperate wooded areas, although it has also been successfully introduced into much more arid areas in Arizona and southern California (Gardner, 1973). Its success must be due in part to its very broad dietary spectrum, as well as to its extremely adaptable behaviour (Hunsaker, 1977).

Didelphis is basically a terrestrial species, and uses the arboreal habitat primarily when searching for food. Literature reports on food preference differ between areas, suggesting that *Didelphis* utilises food which is seasonally or regionally most abundant. For instance, in eastern Texas, Lay (1942) found the stomach contents of *Didelphis* to consist of about 60% animal material (insects, worms, mammals, birds, crayfish and snails) and 40% plant (fruit, green leaves, leaf and log litter, acorns and grass seeds). On the other hand Fitch & Sandidge (1953) found that scats of *Didelphis* in north-eastern Kansas in autumn and winter contained mainly fruit (grapes, hackberries, wild plum and crabapple); other food items were crayfish, insects, corn, rabbit carrion, young snakes, snails, frogs and lizards.

However, it is possible that scat analysis in omnivores could lead to bias towards less digestible food items of plant origin, the more digestible animal material being either completely utilised in the gut or rendered unidentifiable in the faeces. This is supported by the results of Wood (1954) in eastern Texas. He found that of 39 different food items recorded in *Didelphis*, 36 were recorded in the stomach, and only 10 in the faeces.

Notwithstanding, the three studies just described all point to an extremely broad food spectrum in *Didelphis*. The digestive tract is also typically omnivorous in form (Fig. 3.4). The salivary glands are represented by parotids which are smaller than the oval submaxillary glands. The sublingual glands are very small indeed (Flower, 1872). The mucosa of the distal portion of the oesophagus is raised into transverse rugae (Sonntag, 1921). The stomach is simple and globular in shape. The gastric mucosa is largely occupied by fundic glands, the remainder being pyloric glands, with a very narrow zone of cardiac glands at the oesophageal opening (Bensley, 1902). The small intestine is about 250% of body length (Owen, 1868; Flower, 1872), and a collar of lobed glandular tissue (Brunner's glands) is found in the sub-mucosa of adult *Didelphis* immediately distal to the pylorus (Krause & Leeson, 1969). The caecum of *Didelphis* is simple and conical in shape, and about 20–40% of body length. The colon is approximately 150% of body length, has only a loose mesenteric attachment and so is quite mobile (Owen, 1868).

It is unfortunate that while *Didelphis* has been the subject of more general research effort than any Australian marsupial species there has been little emphasis on digestion and nutrition. Most of the research effort in digestive tract physiology in *Didelphis* has been concerned with the functioning of the lower oesophageal sphincter, pyloric sphincter, and ileo-caecal junction muscles. This has been motivated by the fact that smooth muscle arrangements in these regions resemble the situation in man.

Fig. 3.4. Digestive tract of *Didelphis marsupialis*, an American omnivorous marsupial. After Schultz (1976).

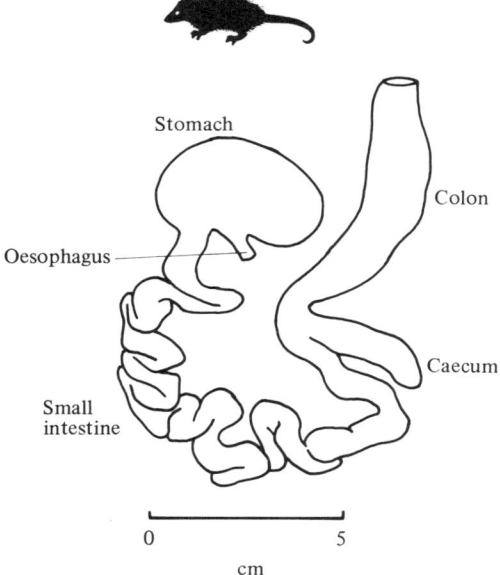

One of the few nutritional experiments with *Didelphis* was conducted by Maller, Clark & Kare (1965). These workers offered *Didelphis* a basal diet of 22.5% protein, 9.3% fat, 42.3% carbohydrate, 21.3% ash and 4.6% water. The digestible energy concentration was then lowered by addition of cellulose, and raised by enrichment with lard. Digestible energy contents thus ranged from 7.1 to 21.3 kJ·g^{-1}. Both dry matter and digestible energy intakes were lowest on the energy diluted diets; on the enriched diets dry matter intake was also depressed, but intake of digestible energy was still higher than on the basal diet. The general conclusion that can be drawn is that, since they did not adjust their dry matter intake very precisely in order to maintain a constant energy intake, factors other than calories, such as palatability, regulate food intake in *Didelphis*.

Dry matter intakes on the basal diet can be calculated to be 38–40 g·kg$^{-0.75}$·day^{-1}, values comparable with intakes by *Trichosurus vulpecula*, the Australian Brushtail Possum, to be described in the next chapter.

Despite earlier indications to the contrary, it now appears that kidney function in *Didelphis* is comparable to that of higher eutherians such as man, cat, dog, sheep, muskrat (*Ondatra*), rabbit (*Oryctolagus*), kangaroo rat (*Dipodomys*) and jerboa (*Jaculus*) in that urinary urea concentration can be raised in response to increased plasma urea levels. The significance of this ability to raise the urinary osmotic ceiling is that it permits the animal to excrete excess solute such as urea without expending any more water. In these animals urea is concentrated more effectively than are electrolytes. That is, urea is excreted in a smaller volume of water than equiosmolar units of electrolytes (Plakke & Pfeiffer, 1970).

Some mammals, however, do not concentrate urea more effectively than electrolytes. These species exhibit a fixed urine osmotic ceiling. Thus the relative concentrations of urine electrolytes and urea vary inversely when this group of eutherians is maximally concentrating urine. These conditions are known to exist in the pig, beaver (*Castor canadensis*), sand rats (*Psammomys*) and the Mountain Beaver (*Aplodontia rufa*) (Plakke & Pfeiffer, 1970). This group must therefore excrete extra urea at the expense of electrolytes (Schmidt-Nielsen & O'Dell, 1961; Schmidt-Nielsen, O'Dell & Osaki, 1961), and as a result requires more water to excrete the total solute load.

The finding that the kidneys of *Didelphis* conform to the first group is supported by its anatomical appearance in that it has clearly defined inner and outer zones in the renal medulla with a relative medullary thickness (Sperber, 1944) intermediate between those of the rabbit and rat (Plakke

& Pfeiffer, 1965). There is a clear association between ability to raise the urinary osmotic ceiling, zonation in the medulla, and relative medullary thickness (Schmidt-Nielsen *et al.*, 1961).

The relative medullary thickness is defined by Sperber (1944) as: 'The medullary thickness × 10 divided by kidney size where kidney size equals the cube root of the dimensions of the kidney.' The relationship between relative medullary thickness and renal concentrating ability for a range of eutherian species is compared with *Didelphis* in Table 3.2.

Omnivorous arboreal marsupials
Diet studies

Four small to medium Australian species are included in this group. They have in common diets consisting of both animal material in the form of arthropods and plant material in the form of some type of exudate high in sugar content. Foliage is not a major food item.

Gymnobelideus leadbeateri (Leadbeater's Possum) is a 150 g arboreal possum found within a restricted region of the central highlands of Victoria where it is most commonly associated with mountain ash (*Eucalyptus regnans*) at altitudes of 600–1500 m. Its diet consists of

Table 3.2. *Comparison of the relative medullary thickness and maximum urine concentration following dehydration*

Species	Relative medullary thickness	Maximum urine concentration ($mOsmoles \cdot kg^{-1}$)
Beaver (*Castor canadensis*)	1.3	520
Pig (*Sus scrofa*)	1.6	1080
Mountain Beaver (*Aplodontia rufa*)	2.9	820
Man (*Homo sapiens*)	3.0	1160
Dog (*Canis*)	4.3	2425
Cat (*Felis*)	4.8	3200
Rabbit (*Oryctolagus cuniculus*)	5.4	1390
Virginia Opossum (*Didelphis virginiana*)	5.7	1497
Rat (*Rattus*)	5.8	3060
Kangaroo-rat (*Dipodomys*)	8.5	6000
Diurnal Sand Rat (*Psammomys obesus*)	10.7	5000

After Plakke & Pfeiffer (1970).

60   3 Bandicoots and other marsupial omnivores

exudates (eucalypt sap, manna and *Acacia* gum) and arthropods (mainly tree crickets, *Apotrechus* sp.) (Smith, 1980).

Petaurus breviceps (the Sugar Glider) (see Plate 3.2) is a 130 g gliding possum which has an extensive distribution throughout much of New Guinea, as well as northern, eastern and south-eastern Australia. It has also been introduced into Tasmania (Smith, 1973). In Australia it is generally found in open (sclerophyll) forest and woodland (Ride, 1970), but also occurs less commonly in tall open forest in the central highlands of Victoria where it overlaps the distribution of *G. leadbeateri* (Smith, 1980). The Sugar Glider's diet consists of *Acacia* gum, eucalypt sap, manna, honeydew and insects (mainly scarabaeid beetles and moths) (Smith, 1978).

Petaurus australis (the Yellow-bellied Glider) is larger (300–450 g), and is distributed along the Australian east coast in tall open forest. It feeds on eucalypt sap, nectar and a variety of arboreal arthropods (Smith & Russell, 1982). Of similar size, *Dactylopsila trivirgata* (the Striped Possum) has a distribution restricted to rainforests and associated sclerophyll forests in north Queensland (Ride, 1970). Its diet has been little studied,

Plate 3.2. *Petaurus breviceps*, the Sugar Glider. (Ray Williams.)

but it appears to feed on social arboreal insects such as ants, termites and bees. The importance of plant sap in its diet is unknown.

Smith (1980) has made a detailed study of the diet and ecology of *G. leadbeateri* and *P. breviceps* in Victoria, together with a survey of diets of arboreal marsupials and the relationships between diet and digestive tract dimensions. Much of the following discussion is taken from this important work.

Several non-foliage plant materials are utilised by *G. leadbeateri*, *P. breviceps* and other omnivorous arboreal marsupial species. All are high in non-structural carbohydrates and so are generally readily digestible. Their protein content, however, is low. Smith (1980) considered that these materials satisfied the bulk of the energy requirements of the omnivorous arboreal marsupials, while arthropods are collected as the major source of protein. The non-foliage plant materials utilised include fruits, nectar, pollen, seeds, saps, manna, honeydew and gums.

Fleshy fruits are generally absent from temperate sclerophyllous eucalypt and *Acacia* forests, and may be low in abundance in cool temperate closed forests in eastern and southern Australia. Tropical closed forests in

Plate 3.3. *Petaurus breviceps* feeding on *Acacia* gum. (Andrew Smith.)

northern Australia and New Guinea may however be rich in fruits. Although most fruiting species in northern Australia have restricted fruiting seasons, in areas of high species diversity fruit would be available throughout much of the year. Fruits may therefore contribute substantially to the diets of ringtail possums (*Pseudocheirus* spp.), cuscuses (*Phalanger* spp.) and tree kangaroos (*Dendrolagus* spp.) in tropical closed forests, although the exact diets of these genera have not been recorded. In some heathland–shrubland communities edible fruits are produced by such plant genera as *Leucopogon*, *Styphelia*, *Astroloma* and *Coprosma*, and may comprise a significant component in the diet of the substantially omnivorous *Cercartetus nanus* (the Eastern Pygmy-possum) (V. Turner, unpublished data).

The value of nectar as an energy resource to small marsupials will depend upon several factors, including seasonality of flowering. Specialisation in nectar exploitation will only be possible if plants have prolonged flowering times or if there is high species diversity of associated plants with staggered flowering times. A number of small marsupials in the genera *Cercartetus*, *Acrobates* and *Petaurus* are known to feed upon nectar. *Tarsipes spencerae* (the Honey Possum), however, is the only marsupial which specialises in the exploitation of nectar, specifically of *Banksia* species (Ride, 1970) (see Plate 3.4).

Pollen grains can have a relatively high protein content (6–28%), which makes them an alternative to insects as a protein source for animals which feed principally upon non-foliage plant materials. However, they are protected by a tough exine coat which is generally difficult to digest (Stanley & Linskins, 1974). Also, pollen production is unlikely to be excessive in comparison with nectar production which may continue throughout the life of the flower. Pollen is thus unlikely to be a principal food item of many animal species. It has been reported in the diets of *Petaurus australis*, *P. breviceps*, *C. nanus* and *T. spencerae*. Although the first three species feed extensively on insects, Duncan (1979) found no insect remains in 16 guts and 8 faecal samples of *T. spencerae*, which suggests that this species may rely upon pollen rather than insects as its primary protein source.

There appear to be no specialist granivores (seed eaters) among marsupials although the omnivorous bandicoots and bilbies (Johnson, 1980) will eat seeds when these are available. Among the arboreal marsupials *Burramys parvus* (the Mountain Pigmy-possum) maintained in captivity have been observed by M. Fleming (Smith, 1980) to extract seeds from mature seed capsules of *Eucalyptus pauciflora* with the aid of its lower incisors, and to make seed caches, giving it the potential to withstand seasonal and temporary seed shortages.

The saps of plants are rich in soluble sugars (Stewart *et al.*, 1973). However, phloem ducts must be mechanically severed, and even then in most species the sap cannot be utilised by sap-sucking animals since it does not flow from the site of injury. The most specialised sap-feeding marsupial

Plate 3.4. *Tarsipes spencerae*, the Honey Possum, feeding on a *Banksia* flower. Note the pollen adhering to its vibrissae. (David Walsh.)

is *P. australis* (the Yellow-bellied Glider) (Wakefield, 1970). It makes a V-shaped incision in the trunks of certain *Eucalyptus* species and licks at the exuding sap. *P. breviceps* chews holes into the trunks of *E. bridgesiana* and feeds upon the exuding sap, which may satisfy a substantial part of its energy requirements (Smith, 1981). Incisions made by these two petaurid species may attract other species such as *Acrobates pygmaeus* (the Feathertail Glider) and the arboreal dasyurid *Phascogale topoatafa*.

Where sap-sucking insects have damaged trees sap may flow from the wound and leave a deposit of white encrusting sugars called manna. Manna production in some eucalypts is substantial and may exceed nectar production at certain times of the year (Paton, 1979). Sap-sucking insects frequently consume large quantities of sap, and secrete excess sugars as honeydew. Honeydew and manna production is seasonal and fluctuates

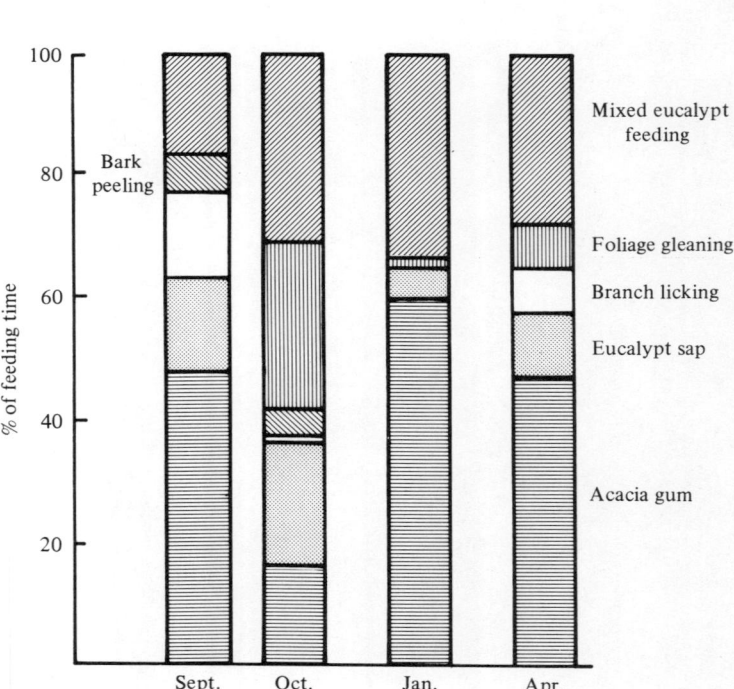

Fig. 3.5. Seasonal changes in the percentage of feeding time spent by *Petaurus breviceps* in six different feeding activities. After Smith (1980).

with insect population levels. These products of insect attack may be seasonally important in the diets of *P. australis, P. breviceps, G. leadbeateri* and *A. pygmaeus* (Smith, 1980).

Gums, like manna, may be produced in response to insect, bird or other mechanical damage to certain trees, principally *Acacia* species. They are produced mainly during the hottest and driest periods of the year. Unlike manna, gums are carbohydrate polymers containing pentosans, and are thus much more difficult to digest. In addition, gums may increase the water and sometimes the protein requirements of animals feeding upon them by binding both water and protein and increasing their faecal excretion (Adrian, 1976). *P. breviceps* and *G. leadbeateri* both feed extensively upon *Acacia* gums (see Plate 3.3). *Acacia mearnsii* gum forms a major component of the diet of *P. breviceps* in southern Victoria during winter when insects are low in abundance.

Seasonal changes in the percentage of Sugar Gliders' feeding time spent searching for different food resources (Smith, 1980) are shown in Fig. 3.5.

Gut morphology and function

The digestive systems of the omnivorous arboreal marsupials examined by Smith (1980) are, with three exceptions, quite similar to those of the bandicoots and *Didelphis* already discussed in this chapter, with only modest development of the caecum into a fermentation chamber, and a short colon. The digestive tract of *Acrobates pygmaeus* (Fig. 3.6) is thus

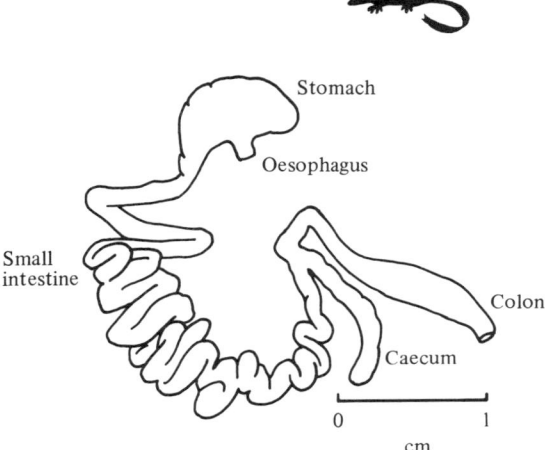

Fig. 3.6. Digestive tract of the tiny *Acrobates pygmaeus*. After Schultz (1976).

66   *3 Bandicoots and other marsupial omnivores*

Fig. 3.7. Digestive tract of *Tarsipes spencerae*. The function of the stomach diverticulum is unclear. After Schultz (1976).

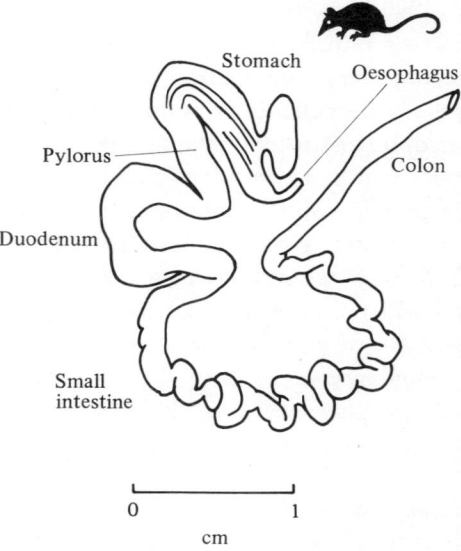

Fig. 3.8. Digestive tract of *Petaurus breviceps*. The large caecum in this omnivore may function as a site for microbial fermentation of *Acacia* gum.

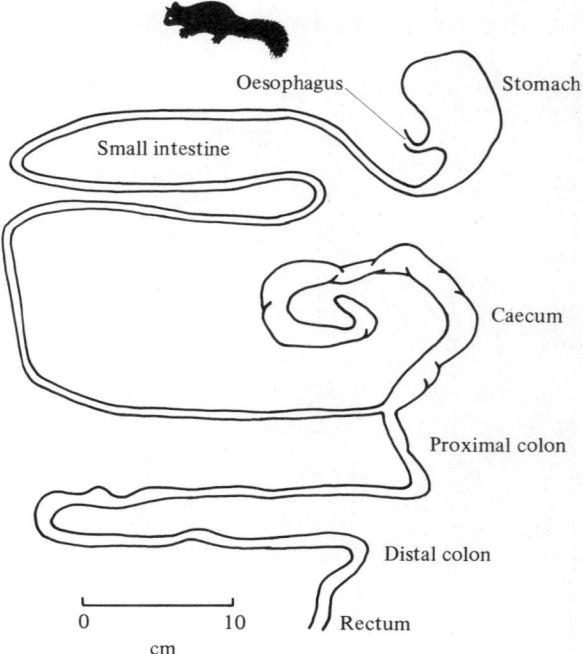

rather similar to those of *Perameles nasuta* (Fig. 3.1) and *Didelphis* (Fig. 3.4). The digestive systems of *P. australis* and *D. trivirgata* also conform to this pattern (Smith, 1980).

The gut of *Tarsipes spencerae* (the Honey Possum) deviates from those so far described in this chapter, not only in the lack of a caecum but also in the two-chambered stomach (Fig. 3.7). The function of this complex organ in *Tarsipes* has not been studied, but presumably the diverticulum serves as a storage organ for nectar. It is also possible that some fermentation of pollen grains may occur in the diverticulum.

The other two exceptions to the general omnivorous digestive system are *P. breviceps* and *G. leadbeateri*, both of which have a large caecum and a long colon. The digestive tract of *P. breviceps* is shown in Fig. 3.8. Both the Sugar Glider and Leadbeater's Possum feed extensively on *Acacia* gums, which, if similar to gums of other tree species, may require bacterial fermentation to be digested. This would then explain the increased development of the caecum and colon of both these marsupial species. Similarly Charles-Dominique (1974) found that several species of lemurs utilised gums, and these animals also possess extremely long caeca (Petter, Schilling & Pariente, 1971).

Smith (1980) attempted to assess the extent to which *P. breviceps* could digest solidified gum from *Acacia mearnsii*, since during winter large particles of undigested gum were frequently found in faeces from Sugar Gliders. Captive gliders were offered *Acacia* gum which had been soaked in water for 12 hours and then broken down into a viscous solution. The apparent digestibility of the dry matter of gum in five *P. breviceps* was 91.5%, indicating that utilisation of the gum should be very high. Whether *A. mearnsii* gum is as digestible as this aqueous preparation at all times of the year is not clear. The appearance of large particles in Sugar Glider faeces in winter and its low water solubility suggest that it is not.

In contrast, gums of *A. dealbata* and *A. obliquinervia*, known to be eaten by Leadbeater's Possums (and presumably also by Sugar Gliders in the same habitat) are readily soluble in water. Thus they may be more digestible than *A. mearnsii* gum. However, because of their high solubility they tend to be rapidly washed off trees, and hence do not provide a year-round food supply for Leadbeater's Possums in the same way that *A. mearnsii* gum does for the Sugar Glider.

Conclusions

Three groups of marsupial omnivores have been considered with respect to their dietary habits and digestive tract morphology. Most species exhibit a larger caecum and longer colon than do carnivores. Only in

Tarsipes spencerae is there any significant enlargement of the stomach. In *Petaurus breviceps* and *Gymnobelideus leadbeateri* the caecum is unusually large for an omnivore. This is explained in terms of their diet which consists partly, and in some seasons of the year, largely of *Acacia* gums. These gums are more difficult to digest than other plant exudates, and their utilisation is thought to be based upon bacterial fermentation in an enlarged caecum. The role of bacterial fermentation in the utilisation of plant foliage in the hindgut of marsupials will be examined in the next chapter.

4
Herbivorous marsupials – the non-macropodids

It is among the herbivores that we find the greatest variations in digestive tract anatomy, and some unusual adaptations to diet. Herbivorous marsupials can be divided into two groups on the basis of the principal site of microbial fermentation in their gut. The macropodids, all of which have their major fermentation region in the foregut, will be dealt with in the next chapter. In this chapter we will consider the non-macropodid herbivores, in which fermentation is restricted to the hindgut. It should be remembered, however, that all foregut fermenters have a secondary fermentation region in the hindgut. This appears to be so among both the Eutheria and the Metatheria. For this and other reasons Hume & Warner (1980), in their discussion of the evolution of herbivory, considered that hindgut fermentation was the more primitive strategy.

Hindgut fermentation

Development of the hindgut of herbivores can involve the caecum alone, the colon, or both. Hume & Warner (1980) referred to two alternative strategies among hindgut fermenters, viz. 'colon fermentation' and 'caecum fermentation'. In the former the proximal colon and the caecum, if present, apparently form one functional unit, without any sorting of particle sizes between the two regions. In 'caecum fermenters', however, the caecum is the primary site of microbial activity, and there is a mechanism for selective retention of small particles in the caecum. Among the Eutheria there seems to be some correlation with the size of the animal. All large (over about 50 kg) hindgut fermenters appear to be colon fermenters (e.g. the Equidae), while all small ones (under about 5 kg) that make substantial use of fibrous foods appear to be caecum fermenters (e.g. the Lagomorpha); at intermediate body weights either strategy may occur. However, many eutherians, particularly among the Rodentia and Primates, are only now being examined. There is also currently much interest in the hindgut fermentation of herbivorous marsupials, particularly folivorous species which utilise *Eucalyptus* foliage as their principal or even sole source of nutrients.

A major hindgut fermentation chamber is found in all members of the families Vombatidae, Phascolarctidae and Phalangeridae, and in some members of the family Petauridae, namely the Greater Glider (*Petauroides volans*) and the ringtail possums (*Pseudocheirus* spp.).

Wombats

There are two genera of wombats found in Australia, the Common Wombat (*Vombatus*) of the forested country of eastern Australia and the Hairy-nosed Wombats (*Lasiorhinus*) of the more arid inland country. Both are large animals, 35–45 kg in body weight. Their teeth are highly adapted to herbivory. They have only a single pair of upper and lower incisors, which together with the other teeth are rootless and grow continuously throughout the animal's life (Troughton, 1965). These characteristics are shared with rodents but are unique among marsupials. Both the incisors and molars of the wombat wear in such a way as to maintain extremely sharp shearing faces on the buccal side of the lower molars and on the

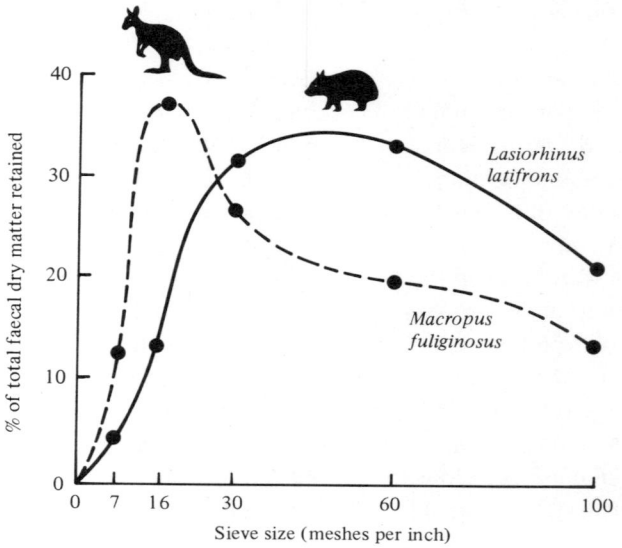

Fig. 4.1. Particle size distribution in faeces from *Lasiorhinus latifrons* and *Macropus fuliginosus* grazing semi-arid grasslands near Blanchetown, South Australia. After Wells (1973).

lingual side of the upper molars. This efficient dental mill is used to break up its highly fibrous plant food, principally grasses (Wells, 1968; McIlroy, 1973). The effect of the efficient dental mill of wombats is seen in the comparison of particle size distribution in faeces collected from *L. latifrons* and the Western Grey Kangaroo (*Macropus fuliginosus*) grazing in the same semi-arid area near Blanchetown, South Australia (Fig. 4.1).

The fine comminution of food by the wombat, as suggested by the greater percentage of fine particles in its faeces, is probably an important factor in its ability to utilise coarse, high-fibre grass. According to McIlroy (1973) the distribution of *Vombatus* in mainland Australia, Tasmania and the islands of Bass Strait is governed largely by the presence of favourable microclimates, suitable burrow conditions and the availability of grasses such as *Poa* spp., *Danthonia* spp. and *Themeda australis*. All these grasses are noted for their tough, highly fibrous nature. Similarly, Wells (1973) found that for the greater part of his two-year study of *Lasiorhinus latifrons* the pasture near Blanchetown contained more than 40% acid-detergent fibre.

The digestive tract of the Vombatidae was the subject of much discussion among the early anatomists. The tract of *V. ursinus* is shown in Plate 4.1.

The most striking morphological feature is the great development of the colon compared with that of the omnivorous marsupials described in Chapter 3. However it is the stomach and the caecum which have attracted most attention. The first reported observations on the stomach of the wombat were those of Home (1808). Although the stomach is small and its external appearance simple, internally the mucosa of the lesser curvature near the cardia is organised into a specialised cardio-gastric gland region. Similar anatomical specialisations have been found among the Marsupialia in the Koala, and among the Eutheria in the Beaver (*Castor*).

The cardio-gastric gland of the wombat is distinctive because of its complex group of mucosal sacculations which open into the stomach lumen via 25 or 30 large crater-like ostia (Hingson & Milton, 1968). These authors could find no histological or cytological difference between the cardio-gastric gland region and the rest of the stomach mucosa, and confirmed the presence of chief cells in the basal region of the gastric glands, parietal cells mainly in the middle 60% of the glands, and neck mucous cells. On this basis they agreed with earlier workers such as Oppel (1896) and Johnstone (1899) that the cardio-gastric gland of the wombat was not a separate organ from the rest of the stomach.

The functional significance of this anatomical specialisation remains

4 Herbivorous marsupials – the non-macropodids

Plate 4.1. The digestive tract of *Vombatus ursinus*.

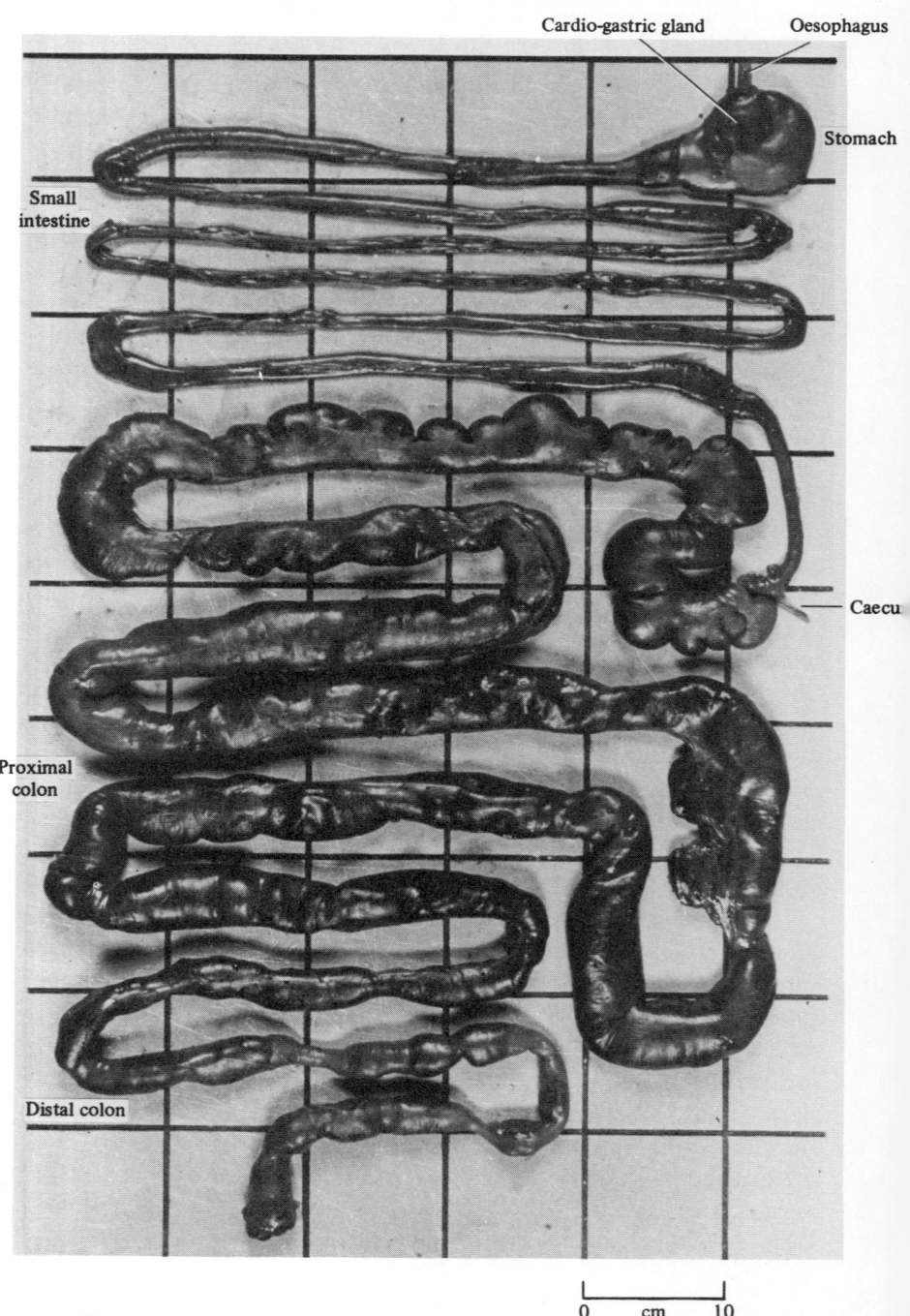

obscure. Oppel (1896) considered that the glandular apparatus of both the wombat and Koala developed into the cardio-gastric gland in order to facilitate the assimilation of large amounts of food, but we now know that the voluntary food intake of both the wombat (Wells, 1973) and the Koala (Cork, 1981) is, if anything, lower than that of most other marsupials.

The caecum was referred to by Owen (1868) as 'extremely short, but wide', and 'provided with a vermiform appendage'. Flower (1872) shared this view of the wombat caecum, but in referring to the vermiform appendage wondered if it was 'as in the higher primates, a remnant of an originally elongated apex of the true caecum'. Lönnberg (1902) and Mitchell (1916), however, regarded the narrow projection at the end of the ileum as the true caecum, the greater part of which has become transformed into a solid vermiform appendix. The wide caecal pouch of Owen (1868) is then merely one of the haustrations of the proximal colon.

The greater size of the colon is functionally the most important feature of the wombat gut (Plate 4.1). Gowland (1973) measured the capacity of the colon of *V. ursinus* and found that it amounted to 68% of the total gastro-intestinal tract. Direct counts of bacteria in the digesta yielded numbers in the order of 10^{10} per gram in all regions of the gut, but numbers were three to eight times higher in the colon than in the stomach and small intestine. In *L. latifrons* the pH of digesta in the colon ranged from 7.0 to 8.0, compared with 2.5 to 4.0 in the stomach. Although volatile fatty acid (VFA) levels or production rates have not yet been measured, microbial fermentation of fibre is probably confined almost entirely in the Wombat to the proximal and distal colon.

Similarly, the long mean retention times (MRT) of digesta in *L. latifrons* found by Wells (1973) (95–210 hours) are no doubt due principally to a slow passage of food residues through the colon. MRT was calculated by staining hay particles with malachite green and following the appearance of stained particles in the faeces after stained feed was offered to the animals for one hour, and then replaced with the usual ration. Shorter MRT (64–123 hours) were found in an earlier study (Wells, 1968), but feed intakes were higher. In fact there was a close relationship between MRT and dry matter intake when data from the two studies were combined (Fig. 4.2).

The difference in dry matter intake between the two studies was explained by Wells (1973) in terms of changes in ambient temperature; food intake of *L. latifrons* was unusually affected by ambient temperature, as suggested by the close relationship between these parameters in Fig. 4.3.

The other aspect of the Hairy-nosed Wombat's physiology of interest

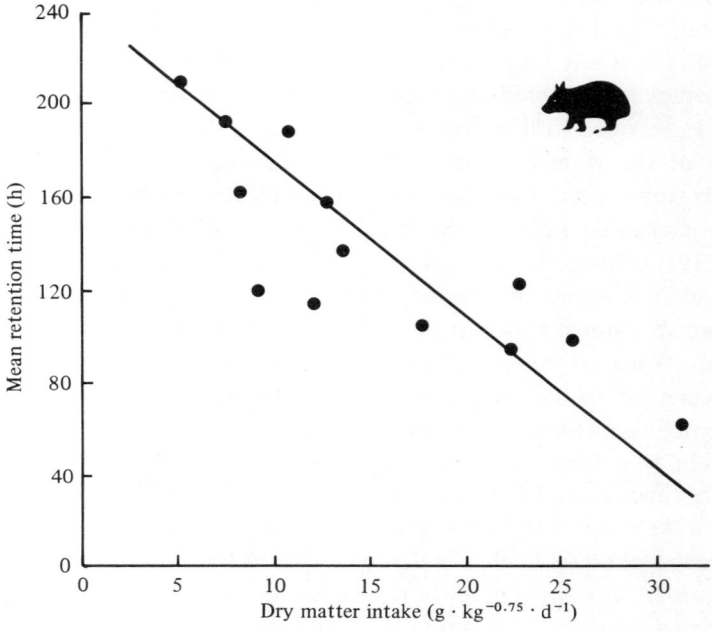

Fig. 4.2. Relationship between mean retention time of stained hay particles and dry matter intake in *Lasiorhinus latifrons*. After Wells (1973).

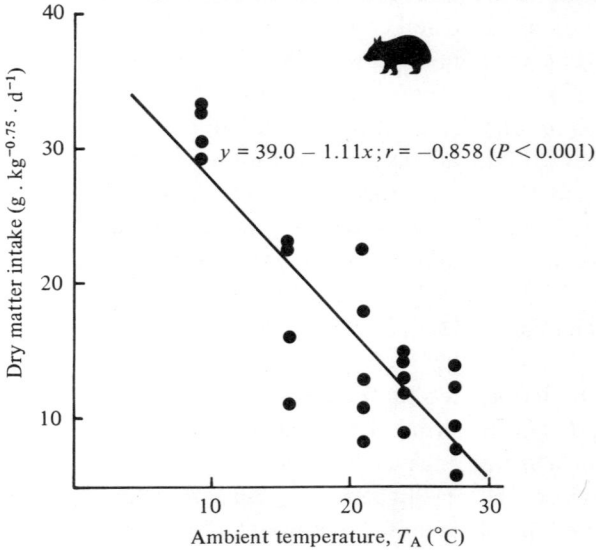

Fig. 4.3. Relationship between dry matter intake and ambient temperature in *Lasiorhinus latifrons*. After Wells (1973).

$y = 39.0 - 1.11x; r = -0.858 \ (P < 0.001)$

to Wells (1973) was its ability to survive on poor-quality pastures without access to free water. One answer to this lies in its behaviour. It is basically nocturnal and fossorial, thereby avoiding extremes of heat (Wells, 1978). In addition, its basal metabolic rate is only 64% of that expected for a marsupial from Dawson & Hulbert's (1970) formula, and only 42% of that expected for a eutherian from Kleiber's (1961) formula (Wells, 1978) (Table 1.3). Related to this is an extremely low water turnover rate of 33.7 ± 9.5 ml·kg$^{-0.80}$·day^{-1} with food and water available *ad libitum* (Wells, 1973). This compares with Denny & Dawson's (1975a) mean among five macropodid species of 90 ml·kg$^{-0.80}$·day^{-1} and Richmond *et al.*'s (1962) mean among seven eutherian species of 134 ml·kg$^{-0.80}$·day^{-1} (Table 1.9). Faecal water loss averaged 38 g per 100 g faeces, compared with 49 and 52 g per 100 g faeces in water-deprived *Notomys alexis* and *N. cervinus* (hopping-mice) (MacMillen & Lee, 1969) and 76 g per 100 g faeces for water-deprived camels (Schmidt-Nielsen, 1964).

Renal water excretion was not as efficient as in the hopping-mice. The kidneys of the Hairy-nosed Wombat were able to concentrate urine to 3100 mOsm per kg water, much lower than the 9370 and 4920 mOsm per kg in *N. alexis* and *N. cervinus* respectively (MacMillen & Lee, 1969), but similar to the value of 2780 mOsm per kg recorded by Dawson & Denny (1969) in the Red Kangaroo (*Macropus rufus*).

Wells (1973) also found that *L. latifrons* was better insulated than a dasyurid or eutherian of similar body weight, and that insensible water loss was low. The combination of all these attributes, together with a labile body temperature (Wells, 1978), helps to explain the water economy of the Hairy-nosed Wombat. Similar studies with the Common Wombat from more mesic habitats would be a valuable addition to our knowledge of adaptations to the environment within the Vombatidae.

The arboreal folivores

This group includes the family Phalangeridae, consisting of the Cuscus (*Phalanger maculatus* and *P. orientalis*), the Brushtail Possums (*Trichosurus* spp.) and the Scaly-tailed Possum (*Wyulda squamicaudata*), the family Phascolarctidae, of which the Koala (*Phascolarctos cinereus*) is the sole member, and the larger members of the family Petauridae, the Ringtail Possums (*Pseudocheirus* spp.) and the Greater Glider (*Petauroides volans*). Other smaller members of the Petauridae eat some foliage, but it is not their principal dietary item.

Much has been written about the feeding habits of the Koala, but most of this is largely anecdotal. Nevertheless the Koala, along with the Greater

Glider, remains the most strictly folivorous of all the marsupials in that its diet consists almost entirely of *Eucalyptus* leaves. In contrast, the Brushtail and Ringtail Possums feed on a wide range of tree species, including leaves, blossom and fruit. Brushtails also spend considerable time feeding on grasses on the ground (Gilmore, 1967; Freeland & Winter, 1975).

The Brushtail Possums
Trichosurus vulpecula

The Common Brushtail Possum (*Trichosurus vulpecula*) is perhaps the most adaptable of all Australian marsupials. It is one of the few species which can be said to have a continental distribution (Troughton, 1965), it has successfully adapted to the presence of European man, in contrast to many other marsupials which have greatly diminished in numbers, several to extinction, and it has successfully adapted to different conditions in New Zealand since its introduction there in 1858. It has also been studied more extensively in the laboratory than has any other marsupial.

The digestive tract. The digestive system of *T. vulpecula* has been described by Lönnberg (1902) and the stomach by Oppel (1896). Simple in form, the stomach is lined mostly by fundic glandular mucosa, with a limited area of pyloric glandular mucosa on and caudal to the deep indenture on the lesser curvature, and a smaller area still of squamous epithelium near the cardia. There is no cardiac glandular mucosa (Fig. 4.4).

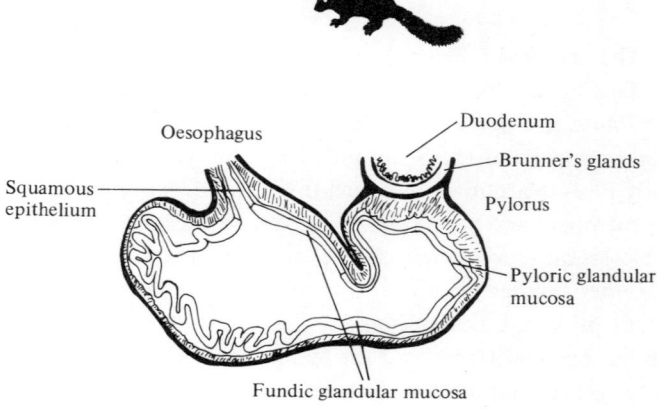

Fig. 4.4. Stomach of *Trichosurus vulpecula*, longitudinal section. After Oppel (1896).

The arboreal folivores

Compared with the simple stomach, the caecum and proximal colon of the Brushtail are well developed (Fig. 4.5). Lönnberg (1902) remarked that 'it will be evident that the caecum of *Trichosurus* is an organ which has become to a considerable degree specialized for digestion (and reabsorption)'. He also noted that although the caecocolic sphincter appeared weak, four other sphincters along the length of the caecum were strongly developed, and that 'there is no doubt that in the living animal they are capable of entirely shutting off one portion of the caecum from the other, thus retaining the enclosed food during a suitable time for decomposition'. However, it is more likely that what Lönnberg (1902) saw, in his preserved specimen, were waves of contraction fixed during their movement along the length of the caecum. In freshly killed animals these caecal contractions continue along the proximal colon as well.

This observation, together with the wide diameter of the proximal colon

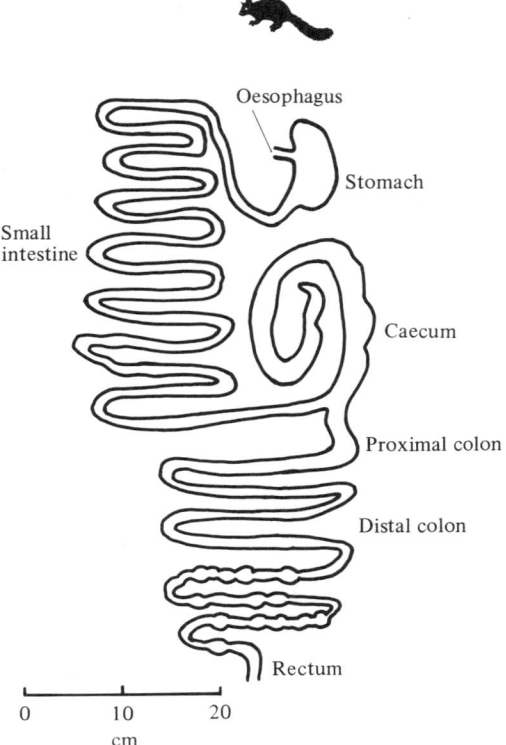

Fig. 4.5. Digestive tract of *Trichosurus vulpecula*. After Harrop & Hume (1980).

(Fig. 4.5), suggests that the digestive function of the caecum of *T. vulpecula* may be shared by this section of the colon also.

Digestive physiology. The first nutritional experiments with *T. vulpecula*, and for any marsupial for that matter, were reported by Honigmann (1936). He measured the rate of passage of digesta through the animals by feeding them wheat grains stained with Loeffler's methylene blue. The stained grain was fed for only a short period, about ten minutes, and then the animals were given access to their regular diet. From the pattern of appearance of the stained grains in the faeces Honigmann (1936) determined that the minimum passage time in *T. vulpecula* was 10 hours and the maximum 96 hours, much longer than in a number of primate species of similar body size that he examined at the same time.

In a further series of experiments Honigmann (1941) fed *T. vulpecula* diets consisting of a mixture of banana and carrot. Five per cent mealworms was included in one experiment. A ten-day collection period followed a seven-day adjustment period. Nitrogen balance was always positive, even on the 11% crude protein diet of bananas and carrot. However, it was on the high crude fibre digestibility on this diet (79–80%) that Honigmann (1941) remarked, concluding that these high values were probably due to the slow passage of food through *T. vulpecula* which he had measured earlier.

The long excretion times in *T. vulpecula* found by Honigmann (1936) were later confirmed by Gilmore (1970) using three different markers, chromic oxide (given to three animals), plastic chips (one animal) and *Eucalyptus* leaf cuticle in two animals fed a diet of carrot, commercial sheep

Table 4.1. *Rate of passage in* Trichosurus vulpecula *measured with three different markers*

	Marker							
	Plastic chips	Chromic oxide				Leaf cuticle		
Animal		A	B	C	Mean	D	E	Mean
First appearance (h)	8	28	32	16	25	32	44	38
Peak concentration (h)	16	32	56	32	40	56	92	74
Final appearance (h)	80	86	120	52	86	86	112	99

After Gilmore (1970).

pellets daily with fresh greens given every second day. However, there appeared to be significant differences in flow characteristics between the markers (Table 4.1).

Although the number of animals used was small, the plastic chips apparently moved faster than the chromic oxide which was given in gelatin capsules, with *Eucalyptus* leaf cuticle moving slowest. It is now known that the rate of movement of a marker is influenced by such things as its specific gravity, particle size and adsorption characteristics (Faichney, 1975). For these reasons many recent studies of digesta flow in domestic ruminants (Faichney, 1975) and in wild animals (Warner, 1981) have been based on the use of inert chemical markers which associate almost exclusively with either the liquid or particulate phase of the digesta. Polyethylene glycol (PEG) has long been used as a liquid phase marker. However, because of analytical difficulties it has been largely replaced by the complex of chromium with ethylenediamine-tetraacetic acid (Cr-EDTA). Analysis is simplified even further by the use of radioactively labelled Cr as [^{51}Cr]EDTA (Downes & McDonald, 1964). Particulate markers currently used include ruthenium complexed with phenanthroline (Ru-Phe), either unlabelled or labelled ([^{103}Ru]Phe) (Tan, Weston & Hogan, 1971), ^{144}Cerium (Ellis & Huston, 1968), and other rare earth metals (Hartnell & Satter, 1979).

Using the dual marker system of [^{51}Cr]EDTA and [^{103}Ru]Phe Wellard & Hume (1981b) measured mean retention times in Brushtails of 64 hours for fluid and 71 hours for particulate digesta. In this study the animals were fed semi-purified diets based on honey, wheat bran, ground oat hulls, purified wood cellulose, casein, salt and minerals and vitamins. The dietary constituents were manipulated to provide diets of low fibre (17% neutral detergent fibre (Van Soest & Wine, 1967)), and high fibre content (41% NDF). The latter fibre level approximated that of mature leaves of *Eucalyptus viminalis*, one of the Brushtail's preferred food trees.

Excretion patterns of animals fed the low fibre diet were too erratic to permit a realistic estimate of mean retention time of digesta to be made, despite regular food intakes. Thus the values given above are for the high fibre diet only. The difference in mean retention times between the two markers was not significant, indicating that there was no differential retention of either phase of digesta in the Brushtail gut.

Surgical removal of the caecum by Wellard & Hume (1981b) made no difference to the extent to which *T. vulpecula* digested dry matter or fibre (Table 4.2). This was surprising in view of the significant capacity of the Brushtail's caecum (Fig. 4.5), and the absence of any obvious increase in the capacity of the proximal colon of caecectomised animals when sacrificed at the end of the experiment. However, information on rate of passage provided the answer. Caecectomy, instead of leading to decreased retention times, actually resulted in mean retention times increasing to 120 hours for fluid and to 125 hours for particles. Apparently this allowed microbial activity in the proximal colon to increase to such an extent that it compensated completely for the lack of a caecum. Increased retention of faecal pellets may also have occurred in the distal colon, again without any increase in size of the organ.

These results suggest that in the Brushtail Possum the caecum functions largely as a simple extension of the proximal colon to form a single fermentation chamber. This is the 'colon fermentation' strategy of Hume & Warner (1980). Although these authors suggested that 'colon fermentation' was probably restricted to animals of large body size because of their relatively low nutrient requirements per unit of body weight, a low metabolic rate in smaller hindgut fermenters such as *T. vulpecula* may effectively extend downward the size range of herbivores able to utilise fibrous diets by this digestive strategy.

Table 4.2. *Intake and digestion in* T. vulpecula *fed semi-purified diets of two fibre contents; values are means ($\pm$ standard errors) of two animals (three replicates each) per treatment*

	Surgical treatment		
	Intact	Sham	Caecectomy
Low fibre (17% NDF)[a] diets			
Body weight (kg)	2.2 ± 0.6	2.0 ± 0.2	1.8 ± 0.1
Dry matter intake ($g\cdot kg^{-0.75}\cdot d^{-1}$)	24 ± 6.9	17 ± 2.4	25 ± 12.6
Apparent digestibility of dry matter (%)	90 ± 5.9	91 ± 1.2	92 ± 1.4
Digestibility of fibre (%)	50 ± 15.7	57 ± 5.1	55 ± 17.7
High fibre (41% NDF) diets			
Body weight (kg)	2.1 ± 0.2	2.3 ± 0.2	1.6 ± 0.2
Dry matter intake ($g\cdot kg^{-0.75}\cdot d^{-1}$)	31 ± 5.1	23 ± 2.8	20 ± 2.3
Apparent digestibility of dry matter (%)	79 ± 2.1	77 ± 4.0	77 ± 0.3
Digestibility of fibre (%)	54 ± 6.2	64 ± 3.5	66 ± 1.8

After Wellard & Hume (1981b).
[a] NDF = neutral-detergent fibre (Van Soest & Wine, 1967).

Metabolism. The low basal metabolic rate of *T. vulpecula* has already been mentioned in Chapter 1. Allied to this is a low maintenance requirement for nitrogen (Wellard & Hume, 1981a) as shown in Table 1.7, and a low field water turnover rate (Kennedy & Heinsohn, 1974) as shown in Table 1.9. Water economy in *T. vulpecula* would also be expected to be maximised by its renal anatomy. The presence of long loops of Henle in the kidney suggests that the Brushtail should be able to form a concentrated urine (Reid, 1977). Deprivation of food and water for three days increased urine osmolality from 184 to 1503 mOsm per kg water, 4.8 times the plasma concentration. Although higher urine osmolalities have been recorded for other marsupial species (up to 3100 mOsm per kg in the Hairy-nosed Wombat), Reid's (1977) results indicate good renal concentrating ability in *T. vulpecula*, and it is likely that a higher osmolality would have been obtained had the animals been deprived of food and water for a longer time, or allowed access to only dry food during the period of water deprivation.

These findings, together with its broad food spectrum and its ability to utilise high fibre diets (Wellard & Hume, 1981b), help to explain the wide distribution of *T. vulpecula* in Australia, and its success following introduction into New Zealand. In Western Australia *T. vulpecula* living within the range of fluoroacetate-containing plants possess an unusually high tolerance to sodium monofluoroacetate (compound 1080), a metabolic poison which blocks the tricarboxylic acid cycle at the citrate stage. Mead, Oliver & King (1979) demonstrated that the Western Australian Brushtail was nearly 150 times more resistant to fluoroacetate intoxication than was the South Australian Brushtail. The high tolerance of the Brushtails in Western Australia has enabled them to exploit the foliage of a wider range of plant species than would otherwise be possible. We will return to the problem of detoxification of plant secondary products when we deal with *Eucalyptus* foliage as a food resource.

Trichosurus caninus

The Mountain Brushtail Possum or Mountain Possum (*T. caninus*) (shown in Plate 4.2) is anatomically similar to *T. vulpecula*, but its ecological requirements are very different (Owen & Thomson, 1965; Tyndale-Biscoe & Calaby, 1975). In contrast to the wide distribution of *T. vulpecula* in all wooded habitats in Australia, *T. caninus* is restricted to the tall open forests of eastern Australia and closed forest in the northern part of its range. Where the two species are sympatric *T. vulpecula* is the less abundant. Within the tall open forests of Victoria *T. caninus* selects a wide range of plant species for food, many of which are found on the forest floor or in the lower storey, whereas *T. vulpecula* in the same forest subsists on foliage of eucalypts, often of a single species (Owen & Thomson, 1965). However, the lower fecundity of *T. caninus* and its high

Plate 4.2. *Trichosurus caninus*, the Mountain Brushtail Possum, from the tall open forests of eastern Australia. (Ray Williams.)

site attachment (Tyndale-Biscoe & Calaby, 1975) make it less resilient and adaptable than *T. vulpecula*.

Whether this entirely explains the very limited distribution of *T. caninus*, or whether there are metabolic differences between the two *Trichosurus* species as well, is not known. Barnett, How & Humphreys (1979a) found that several blood parameters, including plasma glucose and protein concentrations, haemoglobin concentration and red blood cell count, showed greater seasonal variation in *T. caninus* than in sympatric *T. vulpecula* in north-eastern New South Wales. They suggested that the lower seasonal variation in *T. vulpecula* may reflect its ability to ameliorate environment stress and so occupy more diverse habitats, although the effect of dietary differences cannot be ruled out.

Further study of blood parameters of the two species from preferred and peripheral habitats (Barnett, How & Humphreys, 1979b) showed that habitat had a large effect on *T. caninus* but little effect on *T. vulpecula*, again indicative of a more adaptable physiology in the latter species. However, an investigation of the metabolism and nutritional requirements of the two species is needed before any metabolic differences can be evaluated.

The Koala

The Koala (*Phascolarctos cinereus*) (shown in Plate 4.3) is remarkable because of its highly specialised diet and feeding behaviour. Surprisingly, though, few systematic studies of diet selection by Koalas have been made. In excess of 50 eucalypt and less than 10 non-eucalypt species have been reported to be browsed by Koalas to some extent. All reports have in common that although selection for preferred species is disproportionately higher than their relative abundance in the plant community, these species are always relatively common. No reports have been made of Koalas existing in areas in which the preferred tree species were at low densities. Also, there is evidence for strong selection for individual trees within a species. The most commonly reported staple food species are *E. viminalis*, *E. punctata* and *E. tereticornis*, and Koala distributions appear to be related to the distribution of these species (Southwell, 1978). However, several other species may be markedly selected in some localities (e.g. *E. ovata*, *E. camaldulensis*, *E. populnea* and *E. paniculata*), and further information is needed to determine which species are suitable for long-term survival of Koalas.

Digestive tract. The Koala is also remarkable in that its digestive tract (Fig. 4.6) exhibits enormous development of the hindgut, both in the caecum and the proximal colon. Owen (1868) remarked: 'In the koala the caecum and large intestines arrive at their maximum of development', and Mackenzie (1918) stated that in the koala 'we meet with the greatest instance of caecal development in the Mammalia'. Several workers have alluded to the probable importance of fermentative digestion in the caecum and proximal colon of the Koala (e.g. Harrop & Degabriele, 1976), and McKenzie (1978) has made a detailed microscopic study of the caecal epithelium. McKenzie (1978) reported that although the caecal epithelium was basically similar to that of other mammals, the distended intercellular spaces and numerous interdigitations of the lateral plasma membrane of epithelial cells gave the ultrastructural appearance of an epithelium adapted for transporting water in much the same way as the rabbit gall bladder (Kaye, Wheeler, Whitlock & Lane, 1966).

Fig. 4.6. Digestive tract of *Phascolarctos cinereus*. After Harrop & Hume (1980).

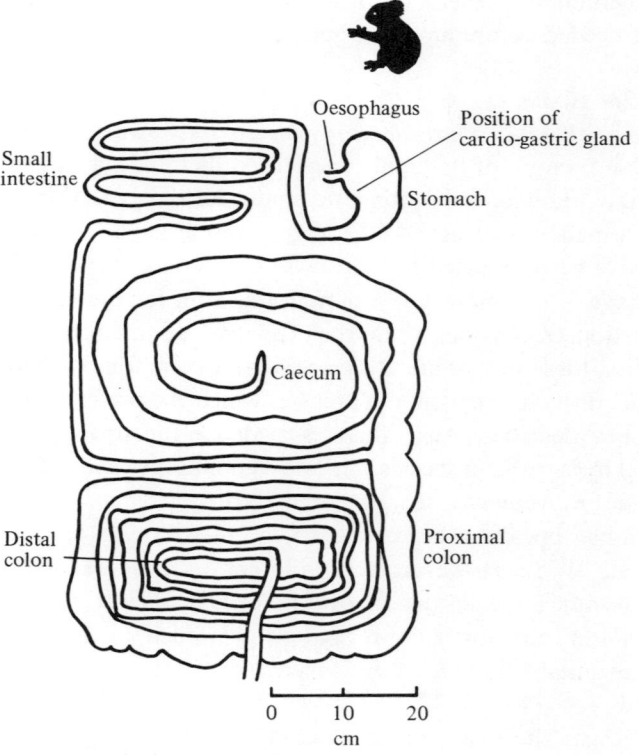

The other feature of the Koala's caecal epithelium was the strong attachment of the microflora with its luminal surface. McKenzie (1978) suggested that this close association of bacteria with epithelial absorptive cells would aid absorption by these cells of the products of microbial fermentation of the eucalypt leaf material of the digesta.

Although the enormous size of the hindgut is the most obvious feature of the Koala digestive tract, the stomach has also received attention from anatomists since, like that of the wombat, it contains a cardio-gastric gland (Oppel, 1896; Johnstone, 1898). The gland measures about 4 cm in diameter and contains about 25 distinct openings (Krause & Leeson, 1973; Harrop & Degabriele, 1976). Although the composition of the gland is similar to that of the wombat, the tubules, which in the wombat are straight and unbranched, in the Koala are branched and thus more complex (Oppel, 1896).

Plate 4.3. *Phascolarctos cinereus*, the Koala, which feeds almost exclusively on *Eucalyptus* foliage. (Ray Williams.)

Nutrition and metabolism. It was not until 1976 that the first nutritional study of the Koala was reported. Harrop & Degabriele (1976) maintained captive Koalas on a sole diet of *E. punctata* foliage throughout the year. Seasonal differences in most nutritional parameters measured were only slight. Dry matter intake averaged 41 $g \cdot kg^{-0.75} \cdot day^{-1}$ in summer and 49 in winter. At the same time apparent digestibility of dry matter declined from a mean of 59.1% in summer to 51.9% in winter. Thus intake of apparently digestible dry matter remained virtually unchanged with season at 25 $g \cdot kg^{-0.75} \cdot day^{-1}$. The above dry matter intakes are generally below those of the macropodid marsupials listed in Table 1.5, with the exception of the Tammar Wallaby (*Macropus eugenii*) (Dellow, 1979), suggesting that the maintenance energy requirement of the Koala, like its basal metabolic rate (Degabriele & Dawson, 1979), is below the general marsupial mean.

In the most complete study available, Cork (1981) measured intakes of digestible energy (DE) and metabolisable energy (ME) by Koalas maintained on *E. punctata* foliage. In line with Harrop & Degabriele's (1976) results on digestible dry matter intakes, there was no consistent seasonal difference in the intake of DE or ME, and all animals remained in slight positive energy balance throughout the year. The mean DE intake was 500 ± 10 $kJ \cdot kg^{-0.75} \cdot day^{-1}$, and the mean ME intake was 430 ± 60 $kJ \cdot kg^{-0.75} \cdot day^{-1}$, compared with the basal metabolic rate of the Koala of 151 $kJ \cdot kg^{-0.75} \cdot day^{-1}$ or 1.75 $W \cdot kg^{-0.75}$ (Degabriele & Dawson, 1979).

Harrop & Degabriele (1976) found that nitrogen balance also remained positive throughout the year, but of course in adult animals it was low. Apparent digestibility of nitrogen, like that of dry matter, was lower in winter than in summer, but the higher food intakes during winter resulted in apparently digestible nitrogen intakes which did not differ significantly with season.

Unfortunately Harrop & Degabriele (1976) were not able to estimate the maintenance nitrogen requirement of Koalas because of the narrow range of leaf nitrogen content (0.92–1.15%). In his more detailed study Cork (1981) was able to estimate that the Koala required 280 mg dietary nitrogen, or 270 mg truly digestible nitrogen $\cdot kg^{-0.75} \cdot day^{-1}$ for maintenance (Table 1.7). This is perhaps higher than would be anticipated from the other data in Table 1.7 and the Koala's low basal metabolic rate. Indeed, urinary losses of nitrogen were low, in line with the low basal metabolic rate. Another reason for the very low urinary nitrogen loss is suggested by a consistently low UR ratio throughout the year.

The UR ratio is the ratio between the concentrations of urea-nitrogen and total nitrogen in the urine, and it has been used by Kinnear & Main (1975) as an index of recycling of urea from the blood to the gut of wild Tammar Wallabies (*Macropus eugenii*). Its ecological significance for Tammars will be discussed in Chapter 6; recycling of urea to the gut means that less is excreted in the urine.

A high UR ratio is indicative of very limited area recycling within the animal. The highest UR ratio recorded by Kinnear & Main (1975) in *M. eugenii* was 0.78 under laboratory conditions in which urea recycling was independently minimised, and 0.74 in the field during the winter wet season when the protein content of available pasture was highest, 23.6%. The lowest UR ratio of 0.32 was recorded during midsummer when plant protein levels were minimal at 6.8%. The mean UR ratio in Cork's (1981) Koalas was 0.36, and the crude protein content of the *E. punctata* leaf eaten by the Koalas was also consistently low, 6.9–9.2%. The consistently low UR ratio suggests that urea recycling in the Koalas was significant in all seasons of the year.

This suggestion was confirmed by Cork (1981) when he measured urea recycling directly using an isotope dilution technique. Of the total amount of urea synthesised in the body each day, 79% was recycled to the gut in winter, 78% in spring, and 76% in summer. This is consistent with a uniformly low protein diet, and is similar to the recycling rate in sheep fed diets of 8% crude protein (Nolan & Stachiw, 1980); sheep fed a 27% crude protein diet recycled only 23% of urea synthesised in the body (Cocimano & Leng, 1967). Compared with eutherian hindgut fermenters such as the horse (Prior *et al.*, 1974) and the Rock Hyrax (*Procavia habessinica*) (Hume *et al.*, 1980) fed low protein diets the urea recycling rate in the Koala was high. This high recycling rate must contribute to the Koala's low urinary loss of nitrogen.

In contrast to the low urinary loss of nitrogen, Cork (1981) found that faecal nitrogen losses were higher than expected. He thought that this was mainly due to a higher metabolic faecal loss of nitrogen (MFN) in the Koala (590 mg per 100 g dry matter intake) than in most other marsupials, both hindgut and foregut fermenters. For example, in the Brushtail Possum Wellard & Hume (1981a) found a MFN value on their semi-purified diet of similar fibre content (41% NDF) of 340 mg per 100 g dry matter intake. The value for the Koala is much more like those reported for ruminants, such as the value of 560 mg per 100 g dry matter intake in the sheep (Harris & Mitchell, 1941).

Part of the reason for the high MFN value for Koalas compared with the Brushtails of Wellard & Hume (1981a) probably lies in the less digestible nature of the fibre of *Eucalyptus* foliage compared with that of the components of the semi-purified diets, and the presence in eucalypt leaves of significant amounts of tannins. Tannins reduce the availability of proteins by complexing with them. This would be expected to affect endogenous as well as dietary proteins, leading to increased endogenous losses.

The less digestible nature of the fibre of eucalypt leaves is reflected in the fact that dry matter intakes by the Brushtails were 30 $g \cdot kg^{-0.75} \cdot day^{-1}$ on the high fibre diets (Wellard & Hume, 1981a) compared with 41 $g \cdot kg^{-0.75} \cdot day^{-1}$ for Cork's (1981) Koalas, despite the lower basal metabolic rate of the Koala (Tables 1.2 and 1.3). Information on the utilisation of *Eucalyptus* foliage by the Brushtail would be a welcome addition to our knowledge of arboreal marsupials, and would allow more direct comparisons to be made with the Koala.

Digestive physiology. The rate of passage of digesta through the Koala is the slowest measured in a marsupial so far. Cork, Warner & Harrop (1977), using the same two markers used by Wellard & Hume (1981b) in the Brushtail, reported an average mean retention time (MRT) of 110 hours (range 68 to 213 hours) for particulate digesta and even longer times, 251 hours (range 96 to 739 hours) for fluid. The pattern of excretion of the markers in the faeces of Koalas and Brushtail Possums after a single oral dose is shown in Fig. 4.7.

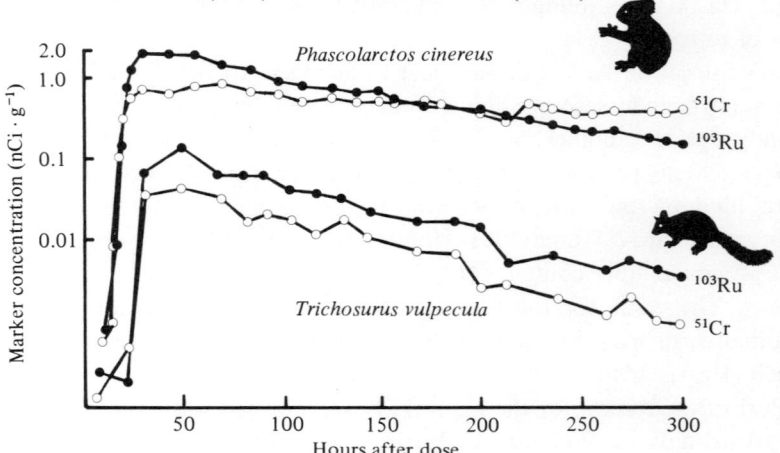

Fig. 4.7. Pattern of excretion of particulate and fluid markers in the faeces of *Phascolarctos cinereus* and *Trichosurus vulpecula*. The rate of passage through the Koala is the slowest measured in a marsupial. After Cork (1981) and Wellard & Hume (1981b).

From the concentrations of the two markers at various sites along the digestive tract of three Koalas killed approximately 3, 12 and 24 hours after receiving a single oral dose of [^{51}Cr]EDTA and [^{103}Ru]Phe, Cork (1981) was able to conclude that while the MRT of particles in the stomach (9.0 hours) was longer than that of fluid (4.5 hours), retention in this organ contributed little to total MRT. It was clear that both phases of digesta accumulated in the caecum and proximal colon, from which, after 24 hours, the particles left at a faster rate. This explains the difference in MRT between fluid and particulate digesta.

The contents of the caecum and proximal colon contained notably higher proportions of fine particles than either the stomach or the distal colon, suggesting that along with fluid, fine digesta particles were selectively retained in the caecum and proximal colon. This is similar to the situation described for the caecum of the rabbit (Björnhag, 1972), but contrasts with the Brushtail Possum in which Wellard & Hume (1981b) found no evidence of differential retention of either digesta phase. Thus the Koala seems to be intermediate between the 'colon fermenters' and 'caecum fermenters' of Hume & Warner (1980); it is also of intermediate body size (5–10 kg). That is, there is selective retention of fluid and fine particles, characteristic of 'caecum fermenters', but this selective retention involves the proximal colon as well as the caecum. Thus particle size distribution in the two regions is uniform, which is characteristic of 'colon fermenters'.

Selective retention of fluid and fine particles would be expected to have several important nutritional consequences for a hindgut fermenter such as the Koala. First, digestibility of the fine particles and soluble constituents of the digesta would be increased (Björnhag & Sperber, 1977). Second, the loss of microbial protein in the faeces may be reduced (Sperber, 1968). Third, although the relatively rapid rate of passage of coarse particles may allow the animal to maintain its intake of digestible dry matter on highly fibrous diets (Hume & Warner, 1980) it should result in lower fibre digestibility.

Cork's (1981) results with Koalas fed only *E. punctata* foliage are consistent with this latter prediction. Although apparent digestibility of the dry matter of the leaves averaged 54.4%, digestibility of the cell wall constituents (NDF) was only 25.1%, of hemicellulose 24.2%, of acid detergent fibre 25.9%, and of cellulose 30.5%. In contrast, apparent digestibility of cell contents (the soluble constituents of the diet) averaged 69.4% (Table 4.3).

Alternatively, it may be argued that even though there is selective retention of fine particles in the Koala caecum/proximal colon, the MRT of coarse particles is still unusually long. Thus the low digestibilities recorded by Cork (1981) for structural carbohydrates may be a reflection of the absolute limit to the digestion of eucalypt leaf by mammalian herbivores, which may be independent of differences in MRT beyond MRTs of 110 hours. A study of the kinetics of digestion of *Eucalyptus* foliage by the use of *in vitro* techniques would be necessary to resolve these alternative explanations for the low digestibility of structural carbohydrates of *Eucalyptus* foliage by Koalas.

Apparent digestibility of lignin was surprisingly high, 18.8%. Lignin is generally regarded as being indigestible by herbivores. The high value for the Koala may be due to the inclusion in the lignin fraction of *Eucalyptus* foliage of some lignin-like phenolic compounds which are degraded by microbial action. These may be in the fine particle rather than the coarse particle fraction of the digesta, and thus be retained in the caecum and proximal colon for extended periods. As can be seen from Table 4.3, apparent digestibility of total phenolics was high, 88–94%; lignin is a polyphenolic compound.

Volatile fatty acid production. Consistent with the finding that fibre digestibility in the Koala was low, Cork & Hume (1980) discovered that volatile fatty acids (VFA) produced in the hindgut fermentation, estimated *in vitro*, contributed perhaps only 9% of the Koala's digestible energy intake (i.e. the energy absorbed).

Table 4.3. *Digestibility of* Eucalyptus punctata *foliage and its fractions by Koalas; values are means ($\pm$ standard errors) of six feeding experiments conducted at different times of the year*

Constituent	Apparent digestibility (%)
Dry matter	54.4 ± 1.2
Cell contents	69.4 ± 0.9
Available carbohydrate	92.1 ± 1.1
Lipid	43.3 ± 2.5
Protein and free amino acids	48.1 ± 2.7
Total phenolics	91.2 ± 0.8
Cell walls (neutral-detergent fibre)	25.1 ± 2.2
Hemicellulose	24.2 ± 5.0
Acid-detergent fibre	25.0 ± 2.0
Cellulose	30.5 ± 3.0
Lignin	18.8 ± 1.7

After Cork (1981).

The *in vitro* technique used by Cork & Hume (1980) to measure VFA production in the hindgut of the Koala, and earlier in the caecum of the Greater Glider (Cork & Hume, 1978), involves removal of digesta from the animal immediately after death. These digesta are then incubated anaerobically in duplicate at the temperature of the gut (37 °C) without any addition of extra substrate or buffer, and samples are removed from the glass incubation vessels at 30-minute intervals for 180 minutes. Total VFA concentrations in the digesta are plotted against time, and the rate of VFA production at zero time (i.e. the zero-time production rate) is estimated by fitting a line of best fit to the data points and extrapolating back to zero time. The results from one of the three Koalas used by Cork & Hume (1980) are shown in Fig. 4.8.

There was found to be no advantage in fitting anything but a straight line to the data points in the case of this or either of the other two animals. This was also the case in six Greater Gliders (Cork & Hume, 1978). In contrast, similar plots of VFA concentration against time with the contents of the forestomach of ruminants (Carroll & Hungate, 1954; Whitelaw, Hyldgaard-Jensen, Reid & Kay, 1970; Hume, 1977a) and macropodid marsupials (Prince, 1976; Hume, 1977a) generally yield curvilinear lines. This is presumably because in the forestomach there is present in the digesta more readily fermentable substrates than in the hindgut. As these substrates

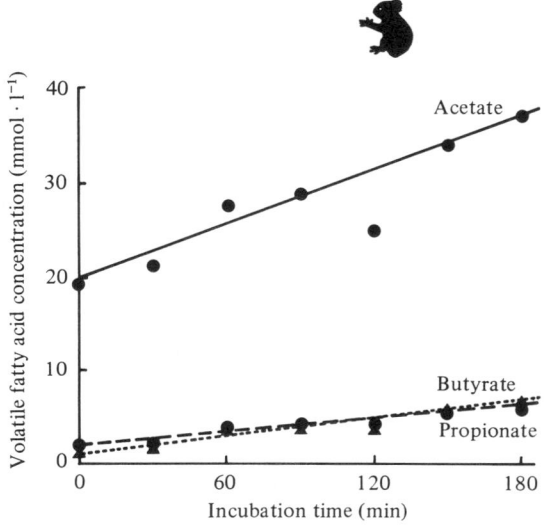

Fig. 4.8. Increase in concentration of volatile fatty acids with time in digesta from the caecum of *Phascolarctos cinereus* when incubated *in vitro*. After Cork (1981).

disappear from the incubation vessel the rate of fermentation tends to decline. In such cases the fitting of a straight line to the data points will lead to an underestimate of the zero-time production rate.

Some readily fermentable substrates, if present, disappear extremely rapidly, and may even have been fermented in the interval between removal of digesta from the animal and removal of the first sample from the incubation vessel. This is thought to be one of the principal reasons why VFA production rates in forestomach digesta measured *in vitro* are consistently lower (by as much as 50%) than estimates made *in vivo* by isotope dilution techniques (Whitelaw et al., 1970).

However, isotope dilution can only be used satisfactorily when it can be assumed that mixing of the injected or infused isotopically labelled VFA with a single large pool of VFA is instantaneous. This situation is approached in the rumen-reticulum of sheep (Leng & Leonard, 1965), but not in the forestomach of macropodine marsupials as we shall see in the next chapter. Nor does it obtain in the hindgut, even though Faichney (1969) was able to achieve reasonable mixing in the ovine caecum in some instances. Thus in hindgut fermenters such as the Koala the choice of method for estimating VFA production rates is restricted to *in vitro* procedures. Fortunately, though, it seems that agreement between *in vitro* and *in vivo* estimates of VFA production is likely to be closer in the hindgut than in the forestomach. This is because of the virtual absence of rapidly fermentable substrates in the caecum and colon.

It has not been feasible to test this in the Koala, but the *in vitro* technique of Faichney (1969) yielded a similar estimate of VFA production rate in the ovine caecum to that obtained using continuous infusion of radioactive acetate *in vivo*. No other studies have been reported in which both *in vivo* and *in vitro* estimates have been made in the hindgut simultaneously. However, on the basis of information available, the assumption that *in vitro* estimates of VFA production rate in the hindgut are a realistic reflection of actual rates of production appears valid.

The second assumption involved in expressing VFA production as a proportion of the animal's intake of digestible energy is that the rate of fermentation measured at one point in time is representative of the rate throughout the remainder of the 24 hours. Again, this has not yet been tested directly in the Koala or the Greater Glider. Certainly in the forestomach both total VFA concentration and the fermentation rate vary markedly throughout the day unless the animal feeds at frequent and regular intervals throughout the whole 24 hours. However, in the caecum of wild or captive rabbits eating *ad libitum* there is little variation in VFA concentration over the 24 hours (Henning & Hird, 1972; Parker & McMillan, 1976). This suggests that the production rate of VFA in the hindgut is fairly constant throughout the day. Thus the estimate of Cork & Hume (1980) that VFA production in the Koala hindgut is equivalent to only about 9% of the animal's intake of digestible energy is probably sound.

Sources of energy. As well as measuring the intake of digestible energy by Koalas fed *E. punctata* foliage in different seasons of the year, Cork (1981) partitioned the total digestible energy intake in winter and spring into five different components (Table 4.4).

The data shown suggest that the most important sources of energy to the Koala are lipids and phenolic compounds. However, phenolic compounds absorbed from the gut of mammals are normally excreted in the urine either directly or following methylation and conjugation with glucuronic or sulphuric acid (McLeod, 1974). Phenolic compounds are obviously present in Koala urine from its tendency to turn black upon exposure to air. Thus the estimate in Table 4.4 of the energy arising from the digestion of total phenolics is an overestimate of the energy available to the Koala from this source.

Similarly, as much as 16% of the Koala's digestible energy intake may be absorbed as essential oils. These potentially toxic plant secondary products are conjugated in much the same way as phenolic compounds and excreted in the urine, or intact oils and their metabolites may be

Table 4.4. *Partitioning of the digestible energy intake of Koalas in winter and spring; values are percentages (mean ± standard error) of total digestible energy intake measured in feeding experiments*

Constituent	% of digestible energy intake	
	Winter	Spring
Protein and free amino acids	6.8 ± 0.5	8.7 ± 0.4
Available carbohydrate	15.0 ± 0.3	14.3 ± 0.5
Lipid	29.8 ± 1.8	19.9 ± 2.1
Total phenolics	32.9 ± 0.6	35.3 ± 1.6
Cell walls	20.7 ± 2.2	17.8 ± 2.0
Total digestible energy intake ($kJ \cdot kg^{-0.75} \cdot d^{-1}$)	533 ± 30	494 ± 20

After Cork (1981).

excreted via the lungs, skin and bile (Eberhard, McNamara, Pearse & Southwell, 1975). Thus the contribution of lipids (which will include the essential oils) to the energy available to the Koala must also be an overestimate.

No information is available on the relative importance of each of the possible excretory routes for essential oils in Koalas. However, it seems likely that the oils, or their detoxification products, could account for a large proportion of the energy lost in Koala urine. Evidence that essential oils and phenolic compounds are excreted in quantity in the urine in Koalas comes from the high ratio of energy lost to nitrogen lost in urine (712 kJ·gN^{-1}) (Cork, 1981). This ratio in ruminants is usually about 40 kJ·gN^{-1} but it can rise to 170 kJ·gN^{-1} when large amounts of hippuric acid are excreted (Black, 1971). The high ratio in Koala urine implies that large amounts of some compound or compounds low in nitrogen content (such as essential oils and phenolics) were excreted.

The conjugation processes involved in the detoxification of essential oils involve an energetic cost to the animal. Eberhard *et al.* (1975) estimated that the Koala normally excretes 1–3 g of glucuronic acid daily. The glucose required to synthesise this glucuronic acid has been calculated by Cork (1981) to be equivalent to as much as 20% of the Koala's fasting glucose requirement, a considerable energetic cost.

Notwithstanding these energy losses, it is apparent from Table 4.4 that the greater part of the energy absorbed and utilised by the Koala is derived from cell contents (i.e. protein and free amino acids, available carbohydrate, lipid and phenolic compounds) rather than from cell walls. This is in line with the small contribution made by VFA production in the hindgut to the energy absorbed by the animal (Cork & Hume, 1980), and with the high activity of several disaccharidases (except trehalase) in the small intestinal mucosa of the Koala (Kerry, 1969) (Table 3.1).

Water metabolism. The water turnover rate measured by Degabriele, Harrop & Dawson (1978) in captive Koalas was quite low in both summer (80 ml·kg$^{-0.80}$·day^{-1}) and winter (92 ml·kg$^{-0.80}$·day^{-1}). These values are only 65–75% of the predicted eutherian level under similar conditions (Richmond *et al.*, 1962). The slightly higher winter value may be related to the higher dry matter intakes of Koalas in that season (Harrop & Degabriele, 1976), and therefore to a greater contribution of metabolic water to the animal's water budget. The animals drank free water in both seasons. Depriving the Koalas of drinking water in winter reduced water turnover to 63 ml·kg$^{-0.80}$·day^{-1}. Under these conditions preformed water in the leaves contributed 58% of the total water input of the animals, and metabolic water the remainder (Table 4.5).

The major avenue of water loss in all cases was evaporation. However, the greatest reduction in water loss when deprived of drinking water was seen in the faecal loss.

When measured in the field, the mean water turnover rate was 179 ml·kg$^{-0.80}$·day^{-1}, approximately double the laboratory values. This difference can be compared with the finding of Hulbert & Gordon (1972) that water turnover in the bandicoot *Isoodon macrourus* in the wild was about 2.5 times that in the laboratory (Hulbert & Dawson, 1974b). There were no significant differences in Degabriele *et al.*'s (1978) study in field water turnover rate in Koalas in three different locations, Magnetic Island near the northern limit of the Koala's range in Queensland, Sydney in New South Wales near the middle of the range, and Philip Island at the southern limit of the range in Victoria (Fig. 4.9).

This suggests that the microhabitat of the Koala is rather constant, at least with respect to water requirements and supply. The high field turnover rate also suggests that water is not limiting to the Koala. Degabriele *et al.* (1978) examined the kidneys from eight Koalas. The relative medullary thickness, which is greater in animals with greater urine-concentrating ability (Schmidt-Nielsen & O'Dell, 1961 and Table 3.2) was only 3.0. By way of comparison the Euro (*Macropus robustus erubescens*), an arid-zone macropodine marsupial, has a relative medullary

Table 4.5. *Partitioning water intake and loss in Koalas under various conditions; values are percentages of total intake or loss*

	Summer Water *ad lib.*	Winter Water *ad lib.*	Winter Water deprivation
Intake			
Water drunk	26	26	0
Preformed water in leaves	45	44	58
Metabolic water	29	30	42
Loss			
Urinary water	20	15	19
Preformed water in faeces	21	25	11
Metabolic water	10	13	9
Evaporative water loss	49	47	61
Total turnover (ml·kg$^{-0.80}$·d^{-1})	80.0	92.3	63.3

After Degabriele, Harrop & Dawson (1978).

thickness of 7.2 (Dawson & Denny, 1969). The ecological significance of this is discussed in Chapter 6. The low value in the Koala indicates that its maximum urine concentration is comparatively low. In their laboratory study Degabriele *et al.* (1978) recorded a maximum urine osmolality of 1692 mOsm per kg water in winter with no drinking water available. Administration of vasopressin (an antidiuretic hormone) induced maximally concentrated urine of 1843 mOsm per kg water, far below values of 9370 and 4920 mOsm per kg water recorded in the desert rodents *Notomys alexis* and *N. cervinus* by MacMillen & Lee (1969).

This, together with the relative proportions of the cortex and medulla given over to filtration (Degabriele *et al.*, 1978) suggests that the Koala kidney is adapted to a habitat with an adequate water supply. When drinking water is not available the Koala takes in water in the leaves it eats, and remains in water balance by producing a reduced volume of a more concentrated urine, by actively reabsorbing urea in the kidney and, most importantly, by reducing markedly the water content of the faeces.

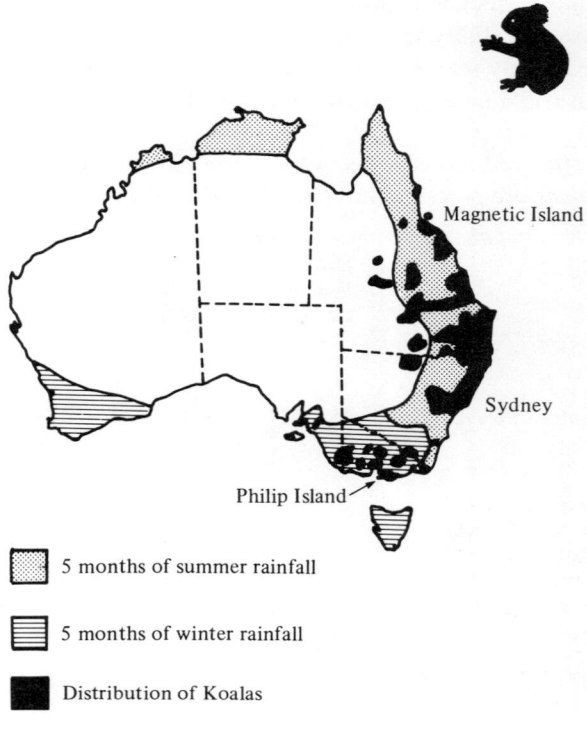

Fig. 4.9. Present distribution of *Phascolarctos cinereus* and the sites of field water metabolism studies. After Degabriele (1977).

The Greater Glider and the Ringtail Possums

Although comparatively little studied, both the Greater Glider (*Petauroides volans* (= *Schoinobates volans*)) and the Ringtail Possum (*Pseudocheirus peregrinus*) are known to be just as strictly herbivorous as the Koala.

The Greater Glider

Petauroides volans (shown in Plate 4.4) has been reported to be a solitary species by Tyndale-Biscoe & Smith (1969a), although they are often seen in pairs during their nightly feeding period. The reproductive rate of Greater Gliders is low and they have a slow population turnover, adaptations to a stable environment (Tyndale-Biscoe & Calaby, 1975). The diet consists principally of leaf-tips, buds and flowers of favoured *Eucalyptus*, including *E. viminalis*, *E. pauciflora*, *E. dalrympleana* and *E. radiata* (Marples, 1973; W. J. Foley, personal communication). However, Troughton (1965) recorded the species in *Casuarina*, and the stomach contents, although finely masticated, appeared to consist of *Casuarina* cladodes. Marples (1973) also recorded bark debris, presumably from *Eucalyptus* in the stomachs of *Petauroides*.

The digestive tract of *Petauroides* includes a simple stomach which can be quite large, and a hindgut dominated by a greatly expanded caecum (Plate 4.5); there is no parallel development of the proximal colon.

Marples (1973) attempted to determine the amounts of food eaten by wild Greater Gliders by weighing the stomach contents of 133 mature animals (i.e. of body weights over 1000 g) which had either been caught by hand during forestry operations or shot while feeding. When the dry weight of stomach contents was expressed as a percentage of body weight and plotted against time of death of the animal it was found that there was a linear loss of material from the stomach during daylight hours and a more rapid rise in content in the evening, starting from 2000 h when the animals begin to emerge from their nest hollows to feed (Fig. 4.10).

The rate of emptying during the day was approximately 1.1 g dry matter·h^{-1}, and the filling rate in the evening 1.7 g dry matter·h^{-1}. By assuming that the rate of emptying during the day was representative of the whole 24 hours (Storr, 1963), the true filling rate was actually 2.8 g dry matter·h^{-1}. Over a feeding period lasting from 2000 h to 0300 h this represents an intake of approximately 20 g dry matter.

However, this must be an underestimate, since W. J. Foley (personal communication) has measured intakes by captive mature Greater Gliders of 46 g dry matter per day. Given the expected greater energy requirements

98   4 *Herbivorous marsupials – the non-macropodids*

Plate 4.4. *Petauroides volans* (*Schoinobates volans*), the Greater Glider, like the Koala is a strict folivore. (Ray Williams.)

The arboreal folivores

Plate 4.5. The digestive tract of *Petauroides volans*.

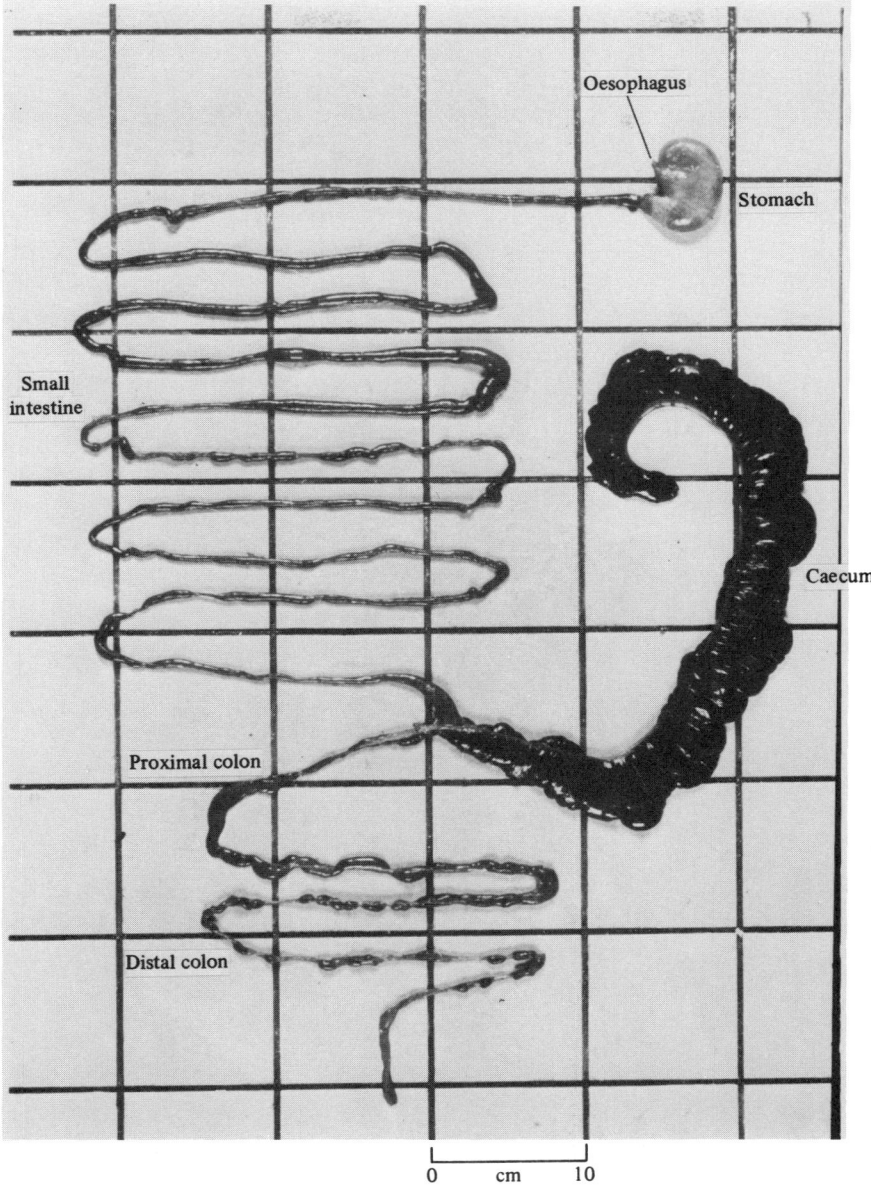

for activity and thermoregulation in the field, intakes of wild Greater Gliders would be expected to be at least as high as those of captive animals.

The estimates made by Cork & Hume (1978) of the rate of VFA production in the caecum of *Petauroides* have previously been alluded to in the discussion in this chapter of the *in vitro* procedure used with both the Greater Glider and the Koala. Cork & Hume (1978) calculated that in the caecum of six Greater Gliders shot while feeding between 2000 h and 2200 h the rate of VFA production was 13–20 μmol·ml^{-1}·h^{-1}. A total daily production of 50 mmole VFA was calculated from the fluid volume of the caecum and by assuming that the hourly rate of VFA production measured was representative of the production rate throughout the day. On the basis of Foley's values for digestible energy intake of captive Greater Gliders, this amounts to approximately 10% of the energy absorbed by the animal. Earlier calculations (Cork & Hume, 1978) yielded values of 17–30% of digestible energy intake, but these relied upon Marple's (1973) estimate of dry matter intake, and hence must be regarded as overestimates.

The apparent digestibility in *Petauroides* of *E. radiata* foliage dry matter is approximately 63%, which is similar to or higher than that in the Koala fed *E. punctata* (52–59%) (Harrop & Degabriele, 1976), and higher than

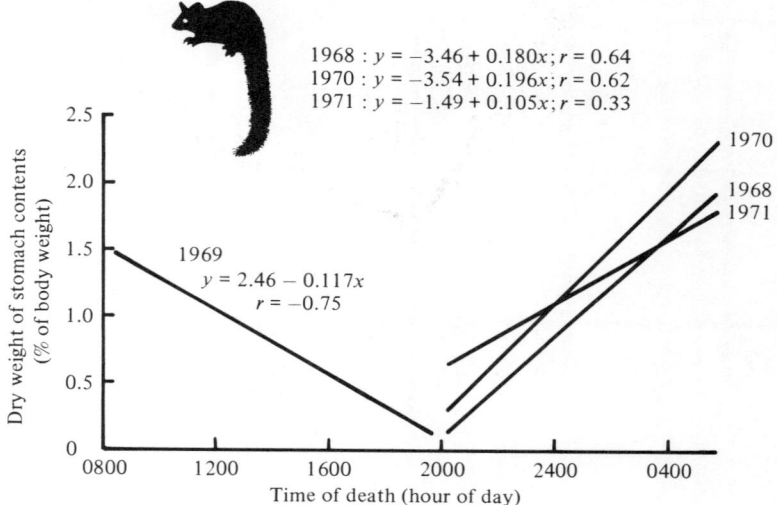

Fig. 4.10. Relationships between dry weight of stomach contents of *Petauroides volans* and time of day. The rate of filling during the feeding period (2000–0300 h) exceeds the rate of disappearance during the non-feeding period. After Marples (1973).

that in three Brushtails maintained on a mixed foliage diet of *E. viminalis*, *E. melliodora* and *Angophora floribunda* (54%) (W. J. Foley, personal communication). Whether these differences in digestibility are characteristic of the marsupial species or are attributable in part to differences between species of eucalypt cannot be judged with our present incomplete state of knowledge.

The other aspect of the digestive physiology of *Petauroides* that has been studied is the rate of passage of digesta. W. J. Foley, using the same dual marker system used by Wellard & Hume (1981b) in the Brushtail Possum and by Cork *et al.* (1977) in the Koala, recorded mean retention times of 45 hours for fluid and 49 hours for particulate digesta. These values are less than those reported by Wellard & Hume (1981b) for the Brushtail Possum fed a semi-purified diet (64 and 71 hours for fluid and particles respectively), but are similar to values recorded by Foley for Brushtails fed the mixed foliage diet described above (51 and 48 hours).

Like the Brushtail Possum, digesta flow in the Greater Glider differs from the Koala in two ways. First, mean retention times are much shorter than those in the Koala, and second, since the mean retention times of fluid and particles do not differ significantly between themselves, there is no selective retention of fluid and fine particles in the caecum. The significance of these differences remains unclear. Much remains to be learned about digestion and metabolism in the Greater Glider before more detailed comparisons with the Koala can be made.

Ringtail Possums

Pseudocheirus peregrinus is widespread throughout south-eastern Australia, ranging from coastal shrublands to the lower edges of snow gum (*E. pauciflora*) communities of the highlands (Thomson & Owen, 1964). It is a socially gregarious species which lives in communal nests or dreys built in the open branches of shrubs and trees, or in mistletoe clumps (*Loranthus* spp.) on eucalypts. Hollows in tree trunks are also used as nest sites. Ringtails have a high reproductive rate and rapid population turnover. In an unstable situation this provides the potential for rapid colonisation or recovery (Tyndale-Biscoe & Calaby, 1975). It is thus a much more resilient species than *Petauroides volans* (Tyndale-Biscoe & Smith, 1969b).

The only dietary study of the Ringtail Possum has been that of Thomson & Owen (1964) in Victoria. These workers claimed *Pseudocheirus* to be apparently one of the strictest herbivores amongst the 'Phalangeridae' (*sic*); in no case were remains of insects or other arthropods other than known ectoparasites found in the stomach contents. Further, no plant species of the field layer were represented, so that, although extensive use is made of ground runways in moving over the home territory, feeding must take place entirely in shrubs or trees.

The diet consists almost entirely of leaves, shoots or flowers, often of only one or two species. For instance, in a climax *Eucalyptus–Kunzea* association, 60% of stomach samples contained *Eucalyptus* fragments, 56% *Kunzea*, 20% *Acacia dealbata* and 4% *Rubus fruticosa* (the introduced blackberry). A single species only was found in 56% of stomachs, 40% contained two species and only 4% contained three species. No seasonal differences in food species were observed (Thomson & Owen, 1964).

In *Eucalyptus* the mature leaf was eaten by *Pseudocheirus* and juvenile foliage avoided, a preference which Thomson & Owen (1964) thought may be correlated with the high concentration of cyanogenic glycosides in the young foliage of certain species of eucalypts; however, this thesis is virtually untested. In contrast, young leaves, shoots and buds of *Kunzea*, *Melaleuca*, *Leptospermum* and young phyllodes of *Acacia dealbata* were preferred to the mature foliage of these species. The seeds of *A. dealbata* were also highly favoured, as also were the fruits and mature leaves of *R. fruticosa*.

The digestive tract of *P. peregrinus* (Fig. 4.11) is similar to that of *Petauroides volans* in that the caecum is well developed, with little enlargement of the proximal colon. Both species must therefore be classified as caecum fermenters in Hume & Warner's (1980) scheme.

Virtually nothing is known of the nutrition or metabolism of Ringtails. Tyndale-Biscoe (1973) reported that M. Griffiths observed soft faeces, quite distinct from the normal excreta, to be reingested by adult Ringtails. At some times of the day the stomach contained fresh leaves, while at other times it contained soft faecal material with a pH of 4.5. It had a high amylase activity and contained protein of bacterial origin. However, Griffiths' observations were actually second-hand (personal communication, 1980), and independent confirmation of these observations is needed before the Ringtail is described as a coprophagous species.

The only other marsupial in which some form of coprophagy has been reported is the Koala. Minchin (1937) observed that young Koalas at the time of weaning ate a semi-liquid faeces taken from the rectum of the mother, which consisted of 'peptonised gum leaves', but which is now thought to be largely contents from the caecum and proximal colon. Sharman (1959) suggested that, apart from its nutritive value (e.g. B vitamins), this may facilitate inoculation of the alimentary tract of the young animal with symbiotic microorganisms.

Eucalyptus foliage as a food resource

From the foregoing discussion on the digestive physiology and nutrition of the Koala, the Brushtail Possums, the Greater Glider and the Ringtail Possums it is apparent that although *Eucalyptus* foliage is an abundant food resource, particularly in the tall open forests and open forests of south-eastern Australia, its utilisation by arboreal folivorous marsupials is potentially limited by two sets of factors. The first of these is a low nutritive value; the second is the presence of toxic secondary products. The relative importance of these two potential limitations to utilisation of *Eucalyptus* foliage is as yet imperfectly understood.

Nutritive value of Eucalyptus foliage

The most complete analyses of the nutrient composition of eucalypt foliage have been those of Cork (1981) and Ullrey, Robinson & Whetter (1981). Earlier analyses were either incomplete or were based on

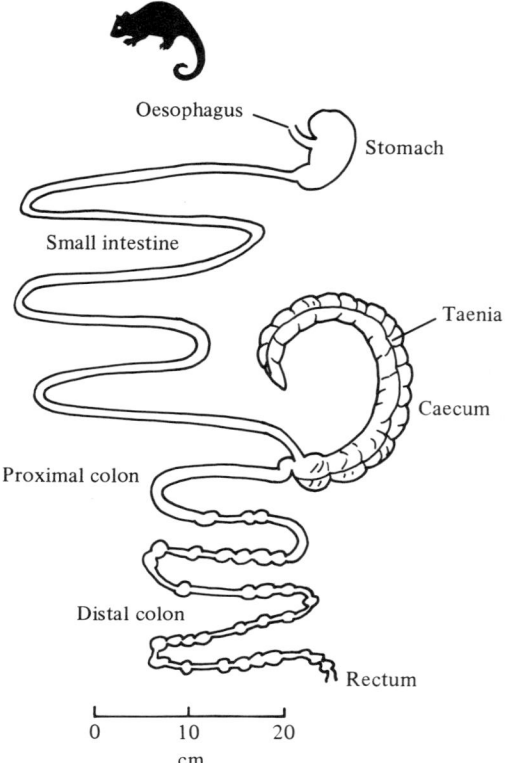

Fig. 4.11. Digestive tract of *Pseudocheirus peregrinus*.

chemical procedures which are now regarded as being of limited value in nutritional studies. The chemical composition of *E. punctata* foliage eaten by Koalas in winter and summer in Cork's (1981) experiments is shown in Table 4.6.

There is little evidence of any seasonal difference in the levels of any of the measured constituents. The diet eaten by the Koalas was thus uniformly low in crude protein and available carboyhdrates (sugars and starch) but high in lipid and phenolic compounds, especially lignin, compared with pasture plant species. Cork's (1981) Koalas selected mainly mature foliage, but this differed little in chemical composition from shoots, with the exception of a much higher lignin content in the mature foliage (Table 4.7). Thus there appears to have been little difference in nutritive value between preferred and rejected foliage. In contrast, Ullrey *et al.* (1981) found at the San Diego Zoo in California that preferred foliage, as compared to rejected foliage, had significantly higher concentrations of crude protein, available carbohydrate, phosphorus and potassium and significantly lower concentrations of lipid, neutral-detergent fibre, acid-detergent fibre, lignin, calcium, iron and selenium. In this case the Koalas selected the younger leaves when offered a mixture of foliage from 11 *Eucalyptus* species, and the younger leaves in general contained higher levels of several important nutrients such as crude protein and phosphorus and lower levels of fibre and lignin.

Table 4.6. *The chemical composition of* Eucalyptus punctata *foliage eaten by Koalas in winter and summer; unless otherwise stated values are expressed as g per 100 g dry matter*

	Winter	Summer
Dry matter (%)	48.0	47.7
Organic matter	97.4	97.4
Crude protein	7.3	9.2
Sugars	9.0	9.8
Starch	1.7	1.5
Crude lipid	19.1	14.0
Total phenolics	22.8	29.3
Cell wall constituents (i.e. neutral-detergent fibre)	35.4	33.2
Hemicellulose	6.5	5.4
Acid-detergent fibre	28.9	27.8
Cellulose	16.0	13.9
Lignin	12.8	13.9

After Cork (1981).

The role of secondary plant metabolites

There is a considerable literature dealing with the effects of secondary plant metabolites on mammalian herbivores and the feeding behaviour and to a lesser extent the metabolic adaptations of herbivores to minimise the adverse effects of browsing foliage containing these potentially toxic compounds (Freeland & Janzen, 1974; Jung, 1977; Swain, 1978). Two types of secondary compounds have been distinguished; those which exert toxic effects directly on the animal's metabolism and those which reduce the potential digestibility of the plant material thereby reducing the animal's fitness (Rhoades & Cates, 1976). Examples of the first type include alkaloids, some amino acids, cyanogenic glycosides and essential oils. The second type includes tannins and lignin.

Freeland & Winter (1975) hypothesised that the density of *Trichosurus vulpecula* populations is indirectly limited by the occurrence of toxic secondary compounds in its major food item, mature *Eucalyptus* foliage. In Freeland & Winter's (1975) study area of open grassy sclerophyll forest, 66% of the Brushtail's feeding times was spent on *Eucalyptus* foliage. They argued from this that Brushtails can eat enough *Eucalyptus* foliage to meet the majority of their energy requirement, but they have to consume non-eucalypt food in order to subsist. According to Freeland & Winter

Table 4.7. *Variation in the chemical composition of foliage of different ages within a single* E. punctata *tree in summer; unless otherwise stated values are expressed as g per 100 g dry matter*

	Shoots	Mature leaves	Old leaves[a]
Dry matter (%)	46.6	49.2	50.8
Organic matter	97.3	97.3	96.3
Crude protein	7.8	8.3	8.1
Sugars	9.9	8.8	8.6
Starch	1.1	1.1	1.2
Crude lipid	10.0	14.5	15.6
Cell wall constituents	13.2	31.9	33.7
Hemicellulose	1.5	4.0	4.7
Acid-detergent fibre	11.7	27.9	29.0
Cellulose	6.7	13.6	14.7
Lignin	4.9	14.2	14.1

After Cork (1981).
[a] Leaves which appeared to be extremely old and tough and which were rejected by Koalas.

(1975) it is the availability of this non-eucalypt food that limits the size of the Brushtail population, not the abundance of *Eucalyptus* foliage.

The secondary compounds most commonly associated with eucalypts are the essential oils. However, polyphenolic compounds, including tannins, are also present in high concentrations in the foliage of some species (Fox & Macauley, 1977). Lignin, which has already been mentioned as being relatively very high in eucalypt leaf (Table 4.6), is also a phenolic polymer. Cyanogenic glycosides are found in high concentration in the young foliage of certain species of eucalypts (Finnemore, Reichard & Large, 1935).

In general, essential oils are those oils which are sufficiently volatile to be removed from plant material by steam distillation. They consist of a variety of compounds, the most abundant of which are related to terpenes (i.e. terpenoids). The terms 'essential' and 'ethereal' refer to the contribution made by these oils to the fragrance of the plant. The demonstration of economically important yields of essential oils from eucalypt foliage (up to 4% of fresh weight) (Baker & Smith, 1920) encouraged naturalists to speculate on the possible relationship between essential oil content and diet selection by Koalas (Troughton, 1965). However, Southwell (1973, 1978) analysed the essential oils of 200 leaf samples from 54 eucalypt species. He found that whether a species or an individual tree had been heavily browsed or avoided was not closely related to total oil yield or to the content of cineole or phellandrene, two of the essential oils thought to be associated with Koala food choice.

Betts (1978) found that a sample of the oil from *Eucalyptus globulus* leaf was composed of the monoterpenoids cineole (60%), α-pinene (11%) and cryptone plus terpinenol (9%) and the sesquiterpenoids erudesmol (14%) and aromadendrene (2%). Variations in the sesquiterpenoid fraction showed some negative correlation with Koala preference for *E. globulus* compared with *E. rudis*, but only 17% of the preference variance could be attributed to the variation in this oil fraction. It must be concluded that no simple relationship exists between essential oils and selection of eucalypt species by Koalas.

The effect of eucalypt essential oils on the gut microorganisms of arboreal folivorous marsupials has not been studied. Essential oils of sagebrush (*Artemesia* spp.) foliage and Douglas fir (*Pseudotsuga menziesii*) needles were shown by Nagy, Steinhoff & Ward (1964) and Oh, Sakai, Jones & Longhurst (1967) respectively to have an inhibitory effect on rumen microorganisms from Black-tailed Deer (*Odocoileus hemionus columbianus*) which had not recently been eating these plant species. However, not all components of the oils were inhibitory, the overall effect depending upon the composition of the oils (Oh et al., 1967). In addition, continuous exposure to such compounds appears to produce a microbial population tolerant of them and which can metabolise them (Oh et al., 1967; Freeland & Janzen, 1974). On this basis it seems likely that the gut microflora of the arboreal folivorous marsupials are probably adapted to eucalypt essential oils. Eberhard et al. (1975) found that no more than 15% of the essential oil ingested by Koalas feeding on E. punctata foliage passed through the gut without transformation, presumbly by the microbiota of the gut, or absorption from the gut.

In eutherian mammals essential oils and other xenobiotics (foreign compounds), once absorbed, are usually handled in a two-phase system of metabolism in the liver. The first step of this process is usually associated with mixed-function oxidase activity (e.g. oxidation, oxygenation, hydroxylation) which introduces a new functional group, giving the xenobiotic a more hydrophilic character. The modified compound is then conjugated with a small molecule (e.g. sulphuric acid, glycine, glucose, glucuronic acid) to make it even more water-soluble and so easily excreted in bile or urine.

Metabolism of the eucalypt terpinoids α-pinene, β-pinene, ρ-cymene and cineole by the Brushtail Possum was shown by Southwell, Flynn & Degabriele (1980) to involve oxidation, hydroxylation and cyclisation. Earlier, Hinks & Bolliger (1957) had observed that the excretion of glucuronic acid in the urine of T. vulpecula more than trebled when the animals were changed from a diet of fruit and vegetable or of Morton Bay fig (Ficus macrophylla) leaves to a eucalypt diet (E. acaciaeformis). Thus from these two pieces of evidence it is apparent that essential oils are probably metabolised by arboreal folivorous marsupials in a manner similar to that demonstrated in eutherians. Glucuronic acid excretion by the Koala has already been mentioned in terms of the energetic cost of the glucose required to synthesise 1–3 g of glucuronic acid excreted each day in the urine (page 94).

Phenolics are the most widespread of the plant secondary compounds. The total phenolic content of different eucalypt species has been reported to vary between 6 and 33% of leaf dry matter (Fox & Macauley, 1977). As with essential oils it is likely that phenolics could affect the animal both via a direct toxicity and via an effect on digestive function. A variable proportion of the total phenolics is in the form of tannins, and the ratio of condensed to hydrolysable tannins is also variable (Fox & Macauley, 1977). Condensed tannins can reduce the availablity of proteins by complexing with them (McLeod, 1974). Fox & Macauley (1977) confirmed the ability of eucalypt tannins to complex protein, but found no relationship between the condensed tannin content of 13 eucalypt species and either feeding rate or nitrogen use efficiency by the larvae of the chrysomelid beetle *Paropsis atomaria*. No studies have been reported of the effects of eucalypt tannins on arboreal marsupials, although Cork (1981) raised the possibility of the relatively high metabolic faecal nitrogen loss in Koalas as being at least partly due to the complexing of eucalypt tannins with endogenous proteins in the gut (page 88).

By injecting [^{14}C]phenol intraperitoneally, Baudinette, Wheldrake, Hewitt & Hawke (1980) were able to examine the metabolism of the simplest phenolic compound in the animals while avoiding any affects on, or by, the gut microflora. Their results are shown in Table 4.8. There was a basic difference between carnivores and folivorous marsupials in the moiety used for the conjugation reaction. The four carnivorous species relied more on sulphate for conjugation, while the two folivores relied more on glucuronic acid. *Petaurus breviceps*, a non-folivorous arboreal omnivore (Chapter 3), resembled the two folivores more than the carnivores, while the two macropodid species were closer to the carnivores. Thus the pattern is not at once completely clear.

Table 4.8. *Urinary metabolities of* [^{14}C]*phenol in carnivorous and folivorous marsupials*

	Percentage of excreted ^{14}C found as:	
	quinol and phenyl glucuronides	quinol and phenyl sulphates
Carnivores		
Sminthopsis crassicaudata	22.5	66.8
Dasyuroides byrnei	5.2	92.5
Dasyurus viverrinus	29.2	72.2
Antechinus swainsonii	26.3	62.5
Terrestrial herbivores		
Potorous tridactylus	18.6	69.6
Macropus eugenii	37.1	64.6
Arboreal omnivore		
Petaurus breviceps	63.1	28.8
Arboreal folivores		
Trichosurus vulpecula	75.8	3.8
Phascolarctos cinereus	94.8	2.3

After Baudinette *et al.* (1980).

Baudinette *et al.* (1980) considered that the reason why the two arboreal folivores show a greater reliance on glucuronic acid for conjugation reactions was likely to be the fact that sulphate groups are limited in diets based on plant material. Although Ullrey *et al.* (1981) did not measure sulphur concentrations in eucalypt leaves, the low levels of other minerals such as calcium, phosphorus, sodium, potassium and magnesium, and the low total ash content of *E. punctata* leaves (Table 4.6; Cork, 1981) suggest that sulphate groups may be even more limiting in *Eucalyptus* foliage than in pasture plant species. However, this does not satisfactorily explain the reliance of the macropodids on sulphate for conjugation.

A more likely explanation for the difference, and one which does produce a clear pattern, is related not to availability of sulphate groups but to availability of glucose, the precursor of glucuronic acid. In both carnivores (Chapter 2) and macropodid marsupials (Chapters 5, 6) very little glucose is absorbed from the gut. Instead, these animals must rely upon the energetically expensive process of gluconeogenesis to provide for their requirements for glucose. It is no doubt metabolically more economical for them to rely upon sulphuric acid for conjugation of xenobiotics rather than on glucuronic acid. In contrast, both groups of arboreal species absorb considerable quantities of glucose from their gut, *Petaurus breviceps* because it eats a diet high in available carbohydrate (Chapter 3; Smith, 1980) and the two arboreal folivores because they are hindgut fermenters rather than forestomach fermenters as are the macropodids. Thus in these two groups glucose availability is not such a problem, and it may be advantageous for them to use glucuronic acid for conjugation reactions rather than sulphate, especially if sulphate availability to the arboreal folivores is likely to be low.

More work is needed on the ability of arboreal folivorous marsupials to metabolise secondary plant products before meaningful relationships between levels of essential oils, phenolics and cyanogenic glycosides and the feeding preferences of various arboreal species for *Eucalyptus* foliage can be expected to emerge.

Summary and conclusions

The non-macropodid herbivorous marsupials are distinguished by their dependence on microbial fermentation in their hindgut for utilising fibrous food. This fibrous food is principally grass in the case of the wombats, and *Eucalyptus* foliage in the case of the arboreal species. In all cases the mean retention time of digesta in the gut has been found to be unusually long, despite the small body size of some of the species, particularly *Petauroides volans* (the Greater Glider) at only one kilogram.

This is not consistent with the general inverse relationship that has been found within the Ruminantia between body size and fermentation rate (and rate of passage), but can perhaps be partly explained by the very low metabolic rates found among the arboreal marsupials (e.g. Koala, cuscus) and in the Hairy-nosed Wombat.

Of the arboreal folivores the Koala has been studied in some detail and we now have good descriptions of the numerous ways in which it is adapted to its arboreal life in *Eucalyptus* forests. Less is known about other arboreal species and their utilisation of foliage diets. The utilisation of *Eucalyptus* foliage is potentially limited by two sets of factors: a low nutritive value relative to pasture plant species; and the presence of plant secondary metabolites. The relative importance of these factors in limiting the utilisation of eucalypt leaves is not yet known. However, by analogy with findings from arboreal primates and other folivorous eutherians it is likely that the folivorous marsupials have evolved mechanisms for metabolising many xenobiotics and that the relatively low nutritive value of *Eucalyptus* foliage is the primary factor limiting its exploitation by arboreal species.

5

Herbivorous marsupials – digestion and metabolism in kangaroos and wallabies

Members of the sub-family Macropodinae (i.e. the kangaroos and wallabies) of the marsupial family Macropodidae are among the most obvious elements of the Australian fauna. There are at present approximately 46 extant species of macropodids, divided between the two sub-families Potoroinae, the rat kangaroos (9 species), and Macropodinae (37 species) (Ride, 1970; Poole, 1979). All are foregut fermenters, but the extent to which each species depends upon microbial fermentation in its forestomach probably varies, depending upon food preferences and the nutritional environment of the habitat. The habitats in which macropodines may be found range from moist forest (e.g. the Red-necked Pademelon, *Thylogale thetis*) to desert (e.g. the Euro, *Macropus robustus erubescens*). In this chapter we will examine some of the adaptations of kangaroos and wallabies in their digestive physiology and their metabolism that have enabled them as a group to exploit such a wide range of environments. Initially the discussion will be on the digestive system of kangaroos and the rate of passage of digesta through the gut. The recent studies of Dellow (1979) on this topic will bring us to the point where we must examine in more detail the form and function of the macropodine stomach, the principal site of microbial fermentation.

Digestion in foregut fermenters

In foregut fermenters, in contrast to hindgut fermenters, ingested food is subjected to microbial attack before exposure to gastric and intestinal enzyme action. This has a number of important consequences for the host animal, some advantageous, some disadvantageous. For instance, most soluble carbohydrates will be fermented in the forestomach, with the production of volatile fatty acids (VFA), principally acetic, propionic and butyric acids, together with methane and carbon dioxide. Since the VFA are utilised less efficiently by the host than are hexoses, fermentation of soluble carbohydrates is a disadvantage. On high fibre diets this is a small price to pay for the advantage of the gain in energy from the fermentation of structural carbohydrates (i.e. cellulose and hemicellulose). Microbial synthesis of vitamins and essential amino acids is another advantage of foregut fermenters feeding on highly fibrous material. These microbial products become available to the host animal after passage to the hindstomach and intestine. Detoxification of plant secondary compounds by the forestomach microorganisms would certainly be expected to enable some foregut fermenters to utilise a wider spectrum of plant material.

Because most of the advantages of foregut fermentation would be nutritionally important when the herbivore was consuming food of high fibre content, Hume & Warner (1980) considered that foregut fermentation probably evolved in regions where adequate quantities of food of not very good nutritive value were available. The rapid radiation of the Bovidae among the Artiodactyla (Janis, 1976) and of the Macropodinae in the Miocene and Pliocene concurrent with the development of widespread grasslands (Hume & Warner, 1980) is consistent with this view. Grasses constitute such a food resource over much of the year, as do tropical forests, now the habitat of tragulids (presumed to be similar to primitive ruminants) and other foregut fermenters such as colobid monkeys and tree sloths.

The macropodine digestive tract

Much of the early descriptive work on the anatomy of the macropodine digestive tract was done by English scientists, principally Home (1808, 1814) and Owen (1834, 1839–47, 1868). Later, other anatomists contributed to the discussion as more material found its way into the museums of Europe. Foremost among them were Schäfer & Williams (1876), Oppel (1896), Lönnberg (1902) and Mitchell (1905, 1916).

Several of these early workers referred to the extensive analogy between the foregut fermentation of the macropodines and the eutherian ruminants. Home (1814) stated: 'The stomach of the Kanguroo [sic] in the peculiarities of its structure, forms an intermediate link between the stomachs of animals which occasionally ruminate; those which have a cuticular reservoir; and those with processes or pouches at their cardiac extremity, the internal membrane of which is more or less glandular.' Owen (1834) took the analogy to ruminants further in commenting on the 'resemblance to the ruminating tribes, to which the kangaroos make so near an approach in the complexity and magnitude of the stomach, and the simplicity of the caecum and colon'. Owen (1839–47) later recorded the presence of a gastric sulcus, which he called an oesophageal groove, in kangaroos. The role of the sulcus in digesta movement through the macropodine stomach will be dealt with later in this chapter.

The work of MacKenzie (1918) marked the end of the first flurry of excitement among anatomists caused by the discovery of the marsupial world. It appeared at that stage that there were only minor differences among macropodine species in stomach structure, and that all were essentially analogous to the ruminant stomach in their structural features. There the matter rested until Moir, Somers & Waring (1956) published their work on the Quokka (*Setonix brachyurus*). This was the first recorded study on the digestive physiology of a macropodine, and, as with the earlier anatomical work, it was used to characterise the Quokka as a 'ruminant-like' herbivore. Since then several studies on the digestive physiology of macropodines have appeared in the literature. At the same time there has been a rejuvenation of interest in macropodines among anatomists. Schultz (1976) published a comprehensive account of the blood supply to the digestive tracts of a wide range of marsupials. Gemmell & Engelhardt (1977) described in detail the morphology and histology of the stomach of the Tammar Wallaby. However, only in the last few years has emphasis been placed on the relationship between structure and function in marsupials, and in the macropodines in particular. Much of the following description draws heavily on the studies of Dellow (1979), Langer, Dellow & Hume (1980) and Richardson (1980).

The digestive tract of the Eastern Grey Kangaroo, *Macropus giganteus* (see Plate 5.1), is shown in Fig. 5.1. In this preparation the mesenteric attachments between segments of the stomach have been left intact, and the stomach is viewed from the ventral surface.

The macropodine stomach is a long, tubular, colon-like organ. It consists of two functionally distinct regions, the enlarged forestomach where microbial fermentation of ingested food occurs, and the hindstomach, the acid-secreting region. The forestomach is regarded by Dellow (1979) as consisting of two regions, the sacciform forestomach and the tubiform forestomach. These two regions are delimited *in situ* by a perpendicular plane running from a permanent ventral fold, adjacent to the cardia, to the dorsal wall. The ventral fold can be seen clearly in the photographs of Langer *et al.* (1980) in animals which, after death, were preserved while suspended by chains in a standing position before dissection. Plate 5.2

shows the fold in the stomach of *Macropus eugenii* during dissection. It cannot be seen in eviscerated specimens, which are invariably dissected free from mesenteric attachment and partially uncoiled during preparation.

Thus the separation of the forestomach into sacciform and tubiform regions is somewhat arbitrary. It is also possible that the permanent ventral

Plate 5.1. *Macropus giganteus*, the Eastern Grey Kangaroo. (Ray Williams.)

fold described by Langer *et al.* (1980) may be an artifact of the preparation *in situ* caused by collapse of parts of the stomach wall not held firmly by mesentery. Notwithstanding, the definition of sacciform and tubiform regions of the forestomach is useful in that it describes the gross form of the two regions; it also has a functional basis, as we shall see presently.

The small and large intestines are both shorter than those of ruminants. Most of the small intestine lies loosely coiled caudal to the sacciform forestomach. The caecum is relatively small and tubular, similar to that of ruminants. Taeniae are present on the caecal wall and extend along the wall of the proximal colon. However, they are not well developed, and the associated haustrations are poorly defined. The lumen of the caecum is confluent with the ascending arm of the proximal colon. A gastrocolic ligament attaches the ascending colon to the tubiform region of the forestomach (Richardson & Wyburn, 1980).

Mitchell (1905) considered that in *Macropus bennetti* (= *M. rufogriseus*) 'exactly opposite the normal caecum there is a second caecum, much shorter than the first, but quite distinct', that 'this relic of paired caeca is quite a common occurrence', and discussed the presence of paired caeca in sloths, anteaters and marsupials in relation to 'a paired condition of the caeca being a primitive mammalian character'. In a later communi-

Fig. 5.1. Digestive tract of *Macropus giganteus*. The mesenteric attachments between segments of the stomach have been left intact in this preparation. After Dellow (1979).

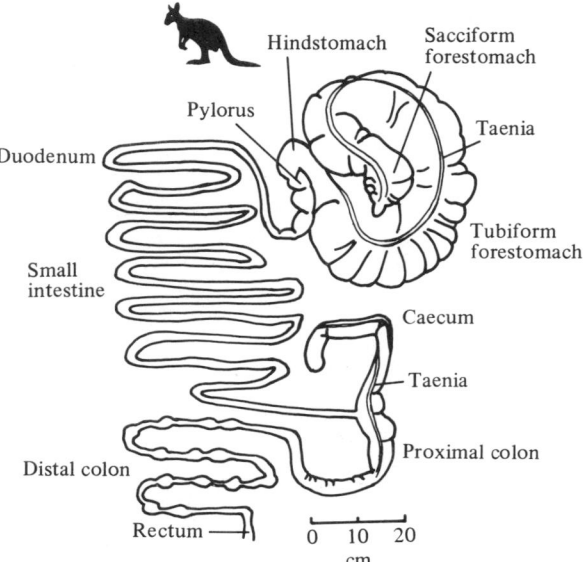

116   5 *Digestion and metabolism in kangaroos*

Plate 5.2.

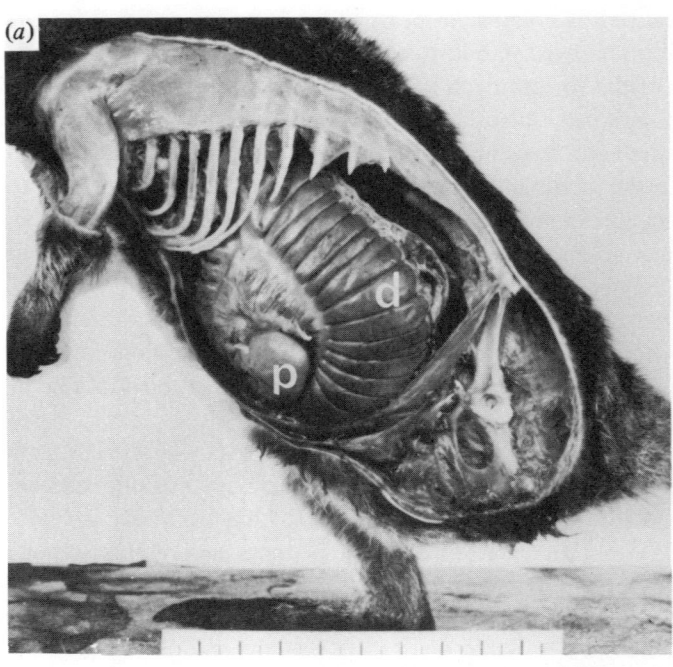

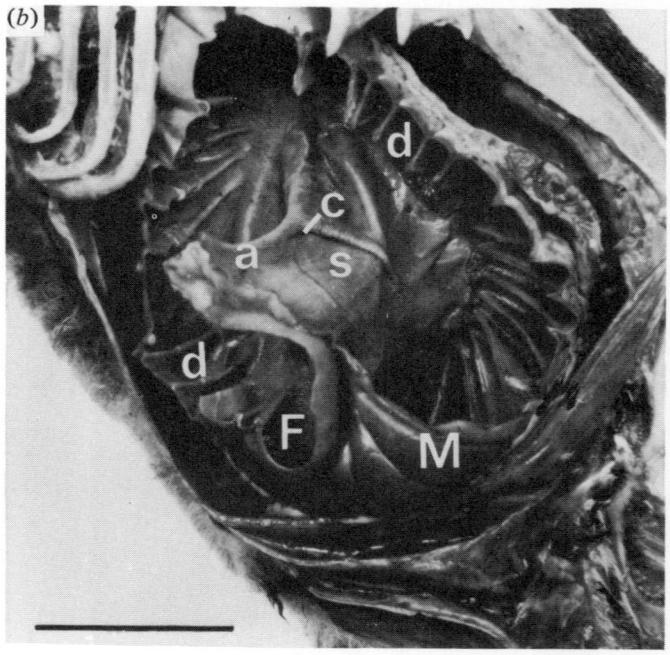

cation (Mitchell, 1916), the large intestine of *Dendrolagus ursinus* was also illustrated with a second, vestigial caecum. The drawings involved are reproduced in Fig. 5.2. However, no evidence of a second caecum has been found in the most recent studies of Dellow (1979) or Richardson & Wyburn (1980).

The distal colon has a small diameter (Mackenzie, 1918), similar to that of the sheep, and is quite short.

Regurgitation

The strong desire by the early anatomists to compare macropodine digestive physiology and metabolism with that of ruminants led them to refer to food regurgitation in the macropodines as rumination. Home (1814) quoted Banks's observations on 'rumination' in the kangaroo. According to Home, the kangaroos kept by Banks ruminated 'when fed on hard food', but conceded that 'it is not however their constant practice, since those kept in Exeter Change have not been detected in that act'. Owen (1834) stated that he had 'more than once observed the act of rumination in the kangaroos. It does not take place while they are erect upon the tripod of their hinder legs and tail. The abdominal muscles are in violent action for a few minutes; the head is a little depressed; and then the cud is chewed by a quick rotatary motion of the jaws. This act was more commonly noticed after physic had been given to the animals, which we may suppose to have interrupted the healthy digestive processes; it by no means takes place with the same frequency as in the true Ruminants.' Rumination was referred to again by Owen in his publications in 1839–47 and 1868.

Wood Jones (1924) also reported on regurgitation in kangaroos, wallabies and bandicoots: 'The animal, after a meal, makes a vigorous heaving movement of its chest and abdomen, and the stomach contents, which are forced up into the mouth appear to be re-swallowed without any further chewing.' He did not refer to the act as rumination. More recently regurgitation has been reported in the Quokka by Moir *et al.* (1965) and Calaby (1958) and in *Macropus rufogriseus* (the Red-necked Wallaby) and *Thylogale billardieri* (the Red-bellied Pademelon) by Mollison (1960). Calaby's (1958) caged Quokkas ejected food boluses at irregular intervals

Plate 5.2. Stages in the dissection of *Macropus eugenii* from the left side. In (*a*) the stomach is shown *in situ.* p, parietal blind sac of the sacciform forestomach; d, haustrations of the tubiform forestomach. The scale line is 20 cm. In (*b*) the permanent ventral fold (a) separating sacciform (F) from tubiform (M) regions of the forestomach can be seen in relation to the cardia (c) and the gastric sulcus (s). d, haustrations. The scale line is 10 cm. From Langer, Dellow & Hume (1980).

which fell through the mesh floor of the cages and could not be recovered by the animals. Thus there are differences in detail, sometimes considerable, in the various descriptions of regurgitation by marsupials, but Barker, Brown & Calaby (1963) concluded that 'it is not analogous to rumination in ruminants, and the term "rumination" should not be used in connection with kangaroos'. They proposed the term 'merycism' (from a Greek work meaning 'chewing the cud'), which is often applied to regurgitation, re-mastication, and re-swallowing in man and animals generally, and which does not have the specialised meaning of rumination as applied to ruminants.

Indeed, the macropodines would appear to have no real requirement for

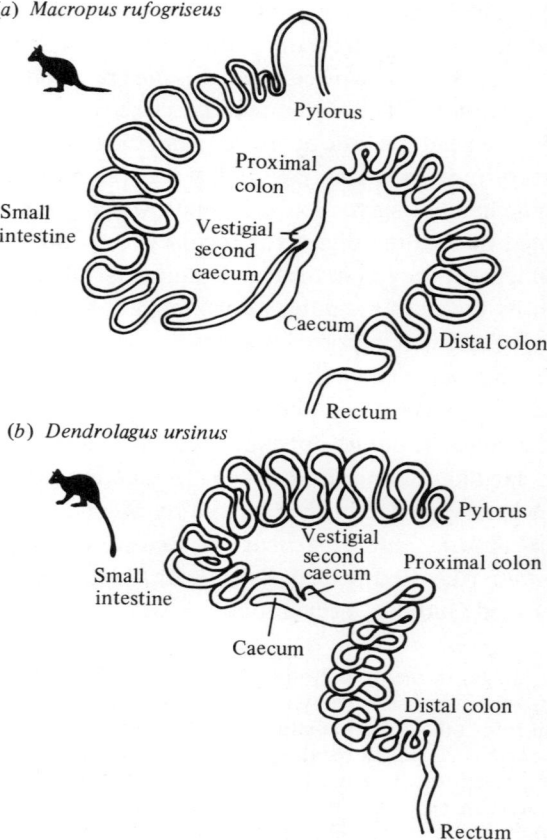

Fig. 5.2. The intestinal tracts of *Macropus rufogriseus* and *Dendrolagus ursinus* showing Mitchell's vestigial caecum. After Mitchell (1905, 1916).

rumination, since they masticate their food more thoroughly initially than do the ruminants, and there is no functional barrier within the stomach analogous to the reticulo-omasal orifice to retain the larger food particles. However, the observation by Dellow (1979) that the frequency of occurrence of merycism can be increased by the addition of crushed grain to a chopped lucerne hay diet suggests that merycism may aid digestion in the macropodines by stimulating salivary secretion. A faster fermentation resulting from grain ingestion would tend to lower digesta pH; a higher saliva flow and greater buffering action in the forestomach would be advantageous under these conditions.

A second type of jaw movement has been observed by a number of workers (Moir *et al.* 1965). It occurs more frequently than merycism, and usually only while the animal is resting, sometimes several hours after feeding. This process does not involve regurgitation of a food bolus, but rhythmic jaw movements continue for up to half an hour. Dellow (1979) has suggested that this also has the effect of stimulating salivary secretion.

Salivary glands

Forbes & Tribe (1969) examined the salivary glands of Red Kangaroos (*Macropus rufus*) and Grey Kangaroos (both *M. giganteus* and *M. fuliginosus*). The parotid gland (see Fig. 5.3) was histologically similar

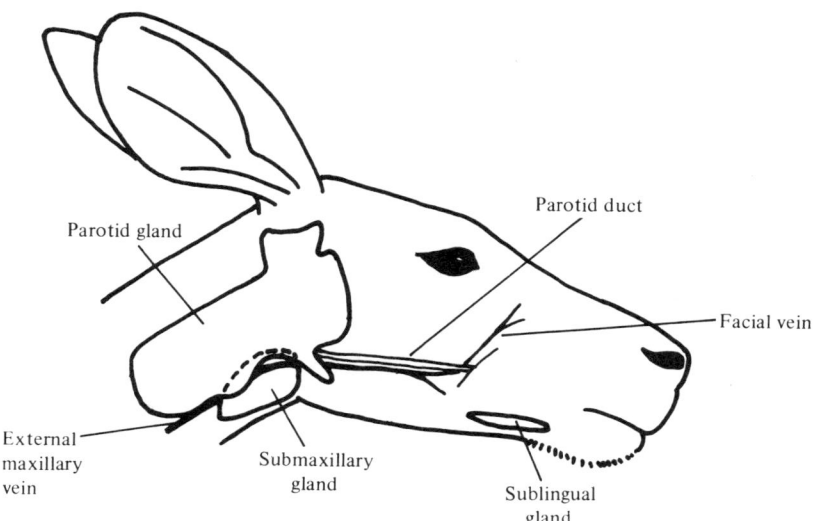

Fig. 5.3. Diagram of the location of the main paired salivary glands of the kangaroo. After Forbes & Tribe (1969).

to that in sheep, but was twice as heavy in relation to body weight. The ovine parotid gland secretes a serous saliva high in buffering capacity in the form of bicarbonate and phosphate ions. In contrast, the submaxillary and sublingual glands secrete a saliva higher in mucus content, but lower in buffering capacity. The same appears to hold for the macropodines, on the basis of data presently available (Table 5.1). The submaxillary gland is of similar size, in relation to body weight, as the submaxillary in sheep (Tribe & Peel, 1963), but has a much greater proportion of serous type secretory cells than the ovine submaxillary.

The other interesting difference between sheep and macropodines is the finding that Red Kangaroo parotid saliva (one animal) contained amylase (Forbes & Tribe, 1969), which is not found in ruminant saliva. Saliva was collected via cannulae inserted into the parotid and submaxillary ducts, so contamination with amylase of bacterial origin is unlikely.

In addition to its buffering role, the high concentration of phosphate in the saliva of both ruminants and macropodines is an important endogenous source of phosphorus for the symbiotic bacteria in the forestomach. Because of their fast growth rates bacteria contain a high ratio of ribonucleic acid-nitrogen to total nitrogen. This means that they have a substantial requirement for phosphate to synthesise the ribose-phosphate chain of the RNA molecule. Barnard (1969) has suggested that ruminants salvage some of this phosphate when bacteria are digested in the abomasum and small intestine. The enzyme ribonuclease occurs at far greater concentrations in the pancreas of ruminants than of non-ruminant eutherians. A close parallel exists among the Marsupialia, at least in those species so far examined. As can be seen in Table 5.2 the concentrations

Table 5.1. *Composition of parotid and submaxillary saliva from the Red Kangaroo, sheep and cattle*

Species	Gland	Ionic concentration (mEq·l^{-1})				
		Na^+	K^+	HCO_3^-	HPO_4^{2-}	Cl^-
Macropus rufus	Parotid	131	16	81	—	20
	Submaxillary	19	15	21	50	16
Ovis aries	Parotid	158–175	10–19	107–142	14–85	4–15
	Submaxillary	15	26	6	54	6
Bos taurus	Parotid	137	14	108	21	15
	Submaxillary	14	14	16	0.5	31

After Forbes & Tribe (1969).

of ribonuclease in the pancreas of three macropodine species approach that found among the Ruminantia, while the concentration in the Virginia Opossum (*Didelphis virginiana*) is low.

Recently A. C. Wilson and his colleagues (personal communication) have found the enzyme lysozyme to be present at high activity in the mucosa of the ruminant abomasum, but not in the stomach of non-ruminant eutherians. Lysozyme hydrolyses the β 1–4 linkages in the polysaccharide chain of bacterial cell walls, and hence probably plays an important role in ruminants in making available to the animal the contents, including RNA, of bacterial cells flowing out of the rumen. Whether lysozyme activity is present in the hindstomach mucosa of macropodids is not known. However, on the basis of other close parallels between macropodines and ruminants, its absence would be surprising.

Table 5.2. *Concentrations of ribonuclease in the pancreas of herbivores*

	Ribonuclease content μg per g wet weight pancreas
Forestomach fermenters	
Red Kangaroo (*Macropus rufus*)	600
Eastern Grey Kangaroo (*M. giganteus*)	530
Tammar Wallaby (*M. eugenii*)	515
Sheep (*Ovis aries*)	1080
Bison (*Bison bison*)	1180
Cow (*Bos taurus*)	1200
Elk (*Cervus canadensis*)	550
Uganda Kob (*Kobus kob*)	270
Hippopotamus (*Hippopotamus amphibius*)	62
Hindgut fermenters with limited forestomach fermentation	
Rat (*Rattus*)	260
Guinea-pig (*Cavia*)	240
Golden Hamster (*Mesocricetus auratus*)	260
Mouse (*Mus*)	395
Hindgut fermenters	
American Opossum (*Didelphis marsupialis*)	20
Horse (*Equus caballus*)	25
Pig (*Sus*)	80
Rabbit (*Oryctolagus cuniculus*)	0.5
Elephant (*Loxodonta africana*)	0.7

After Barnard (1969).

Passage of digesta through the gut

Since microbial digestion of plant structural carbohydrates is a slow process (Hungate, 1966), the extent of fibre digestion will be a function of the residence time of food particles in the gut, particularly the forestomach and/or hindgut. For this reason the rate of passage of stained food particles through the ruminant digestive tract has been extensively studied (Castle, 1956). In ruminants, larger food particles have a longer retention time in the reticulorumen (the forestomach) than do small particles (Blaxter, Graham & Wainman, 1956). This is because particles must be reduced in size by fermentation and rumination before passage out of the rumen via the reticulo-omasal orifice. Thus the size of the stained hay particles could influence the rate of passage measurements, as Castle (1956) illustrated in goats. Nevertheless, useful comparisons between ruminants and macropodines have been carried out with the technique by Foot & Romberg (1965), McIntosh (1966) and Forbes & Tribe (1970).

Data from animals fed chopped lucerne hay (Table 5.3) indicate that, even on a lower voluntary food intake, excretion times in the Quokka, the Red Kangaroo (*Macropus rufus*) and the Eastern Grey Kangaroo (*M. giganteus*), were all shorter than in sheep. Both Calaby (1958) and McIntosh (1966) concluded that this was possibly because of less efficient

Table 5.3. *Dry matter intake and times for excretion of 50% and 90% of stained hay particles in kangaroos and sheep fed chopped lucerne hay ad libitum*

Species	Dry matter intake ($g \cdot kg^{-0.75} \cdot d^{-1}$)	Excretion time (h) 50%	90%	Reference
Setonix brachyurus	47	—	38	Calaby (1958)
Macropus giganteus	49	39	50	Forbes & Tribe (1970)
Macropus rufus	58	35	45	Foot & Romberg (1965)
Macropus rufus	38	41	58	McIntosh (1966)
Macropus rufus	63	28	39	Forbes & Tribe (1970)
Macropodine mean	52	36	48	
Ovis aries	72	41	67	Foot & Romberg (1965)
Ovis aries	64	38	69	McIntosh (1966)
Ovis aries	67	52	89	Forbes & Tribe (1970)
Sheep mean	67	44	75	

mixing of ingested food in the macropodine stomach than in the ovine reticulorumen.

If the results are plotted on a cumulative basis (Fig. 5.4), differences in the pattern of excretion as well as in excretion times become apparent. For example, the time difference between 50% and 90% excretion time was considerably shorter in the macropodines (10–12 hours) than in the sheep (26–37 hours); this is reflected in the steeper cumulative percentage excretion curves of the macropodines.

Dellow (1979) has recently made a more detailed study of digesta flow in macropodines, with some surprising results. By feeding animals a small amount of crushed wheat grain impregnated with [^{51}Cr]EDTA and [^{103}Ru]Phenanthroline, he was able to define the kinetics of flow of both the liquid and particulate digesta in animals fed either chopped lucerne hay or fresh grass *ad libitum*. Additional feed was offered four-hourly to encourage steady state conditions of digesta flow in the gut. The patterns of appearance of the two radioactive markers in the faeces of sheep, the Eastern Grey Kangaroo, the Red-necked Pademelon (*Thylogale thetis*) and the Tammar Wallaby are shown in Fig. 5.5. Cumulative percentage recovery of the markers is shown in Fig. 5.6. Similar patterns were obtained in the Tammar by Warner (1981).

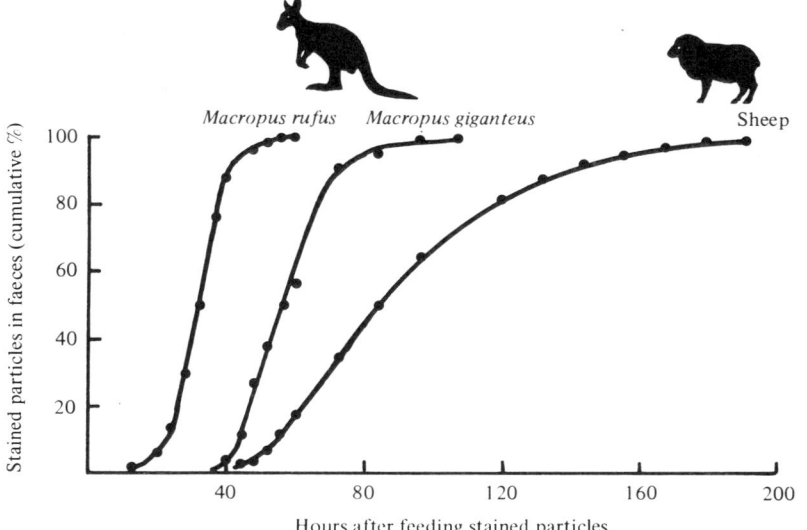

Fig. 5.4. Cumulative rate of passage curves for *Macropus rufus*, *M. giganteus* and sheep fed chopped oaten straw *ad libitum*. After Forbes & Tribe (1970).

The most striking difference is between the sheep and the Eastern Grey Kangaroo, especially in the pattern of appearance of the markers. Separation of the markers during their passage through the ovine digestive tract was minimal (Fig. 5.5). In contrast, there was a distinct separation of the markers in the Eastern Grey Kangaroo, the fluid marker passing through the gut much more rapidly than the particulate marker. The difference in 50% excretion time between the two markers was 16 hours, compared with less than 4 hours in the sheep. This difference can be seen quite clearly in the cumulative excretion curves in Fig. 5.6.

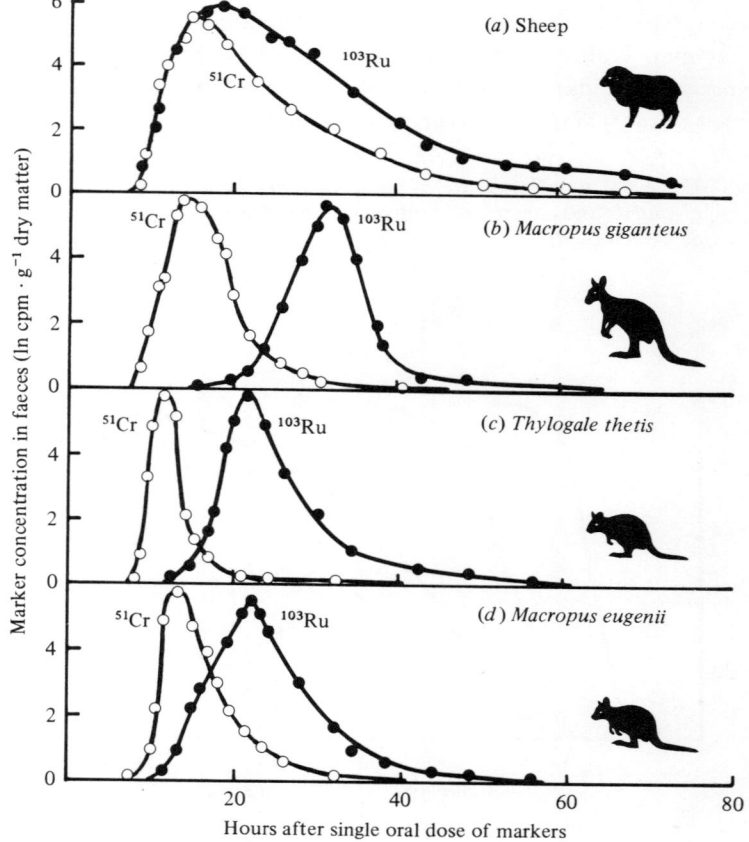

Fig. 5.5. The pattern of appearance of the fluid marker [^{51}Cr]EDTA and the particulate marker [^{103}Ru]Phenanthroline in the faeces after a single oral dose in sheep, *Macropus giganteus*, *Thylogale thetis* and *M. eugenii* fed chopped lucerne hay *ad libitum*. Note (a) the faster elimination of both markers in the macropodine marsupials, and (b) the greater separation of the two markers in the macropodines than in the sheep. After Dellow (1979).

Differences in the patterns of excretion of the markers are also apparent among the three macropodine species. Separation of the markers was not as marked in either the Red-necked Pademelon or the Tammar Wallaby compared with the Eastern Grey Kangaroo. Essentially similar results were obtained for Pademelons and Tammars when fed fresh *Phalaris aquatica* ad libitum.

In a further experiment, Pademelons and Tammars were surgically prepared with an infusion catheter into the hindstomach. When the two markers were injected into the hindstomach the pattern of excretion in the faeces was almost identical for both the fluid and particulate marker (Fig. 5.7). Thus there was no indication of differential flow of digesta through the intestine. Furthermore, excretion times were much shorter than through the whole digestive tract (9 hours versus 23 hours for [^{103}Ru]Phenanthroline).

The conclusions from these results are that:

(a) Retention of digesta in the stomach of macropodines is, like

Fig. 5.6. Cumulative percentage recovery curves for the two markers [^{51}Cr]EDTA and [^{103}Ru]Phenanthroline derived from the curves in Fig. 5.5. The steeper curves for the macropodine marsupials reflect the faster rate of passage of both markers compared with that in sheep. After Dellow (1979).

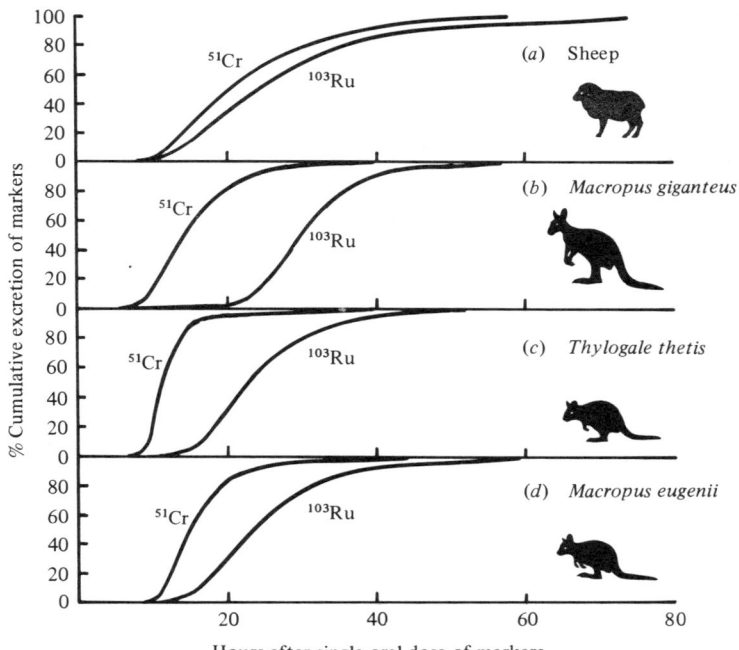

ruminants, much longer than in any other part of the digestive tract.
(b) Separation of the fluid and particulate phases of digesta in the macropodine gut occurs only in the stomach.
(c) In contrast to ruminants, mixing of digesta in the macropodine forestomach does not occur within a single, large pool.
(d) Because of the elongated nature of the macropodine stomach, there must be a significant transit time for both phases of digesta through the forestomach.

Dellow (1979) has described the mode of digesta flow through the macropodine stomach as 'tubular flow'. To understand more clearly the origins of this tubular digesta flow, and of the differences in the marker excretion patterns among the three macropodine species in Fig. 5.5, we must now look more closely at the structure and function of the macropodine stomach.

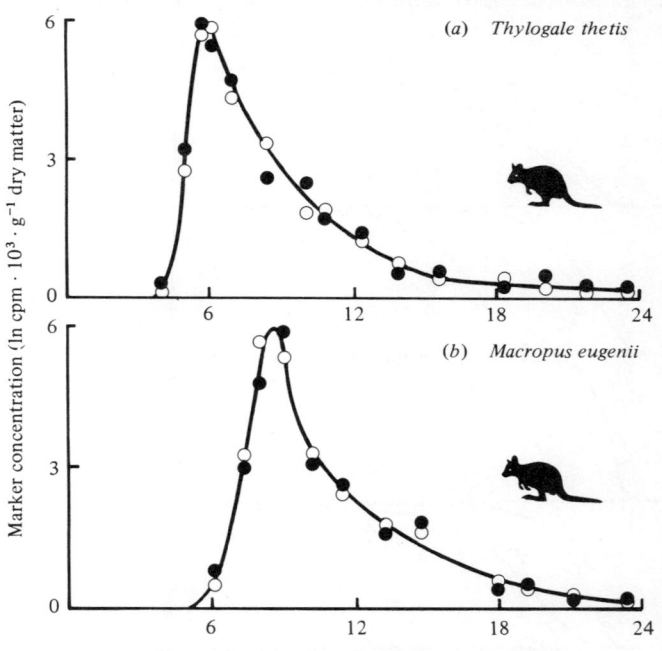

Fig. 5.7. The pattern of appearance of [^{51}Cr]EDTA and [^{103}Ru]Phenanthroline in the faeces of *Thylogale thetis* and *Macropus eugenii* after a single injection of both markers into the hindstomach. There was virtually no separation of the two markers in the intestine and excretion times were very short. After Dellow (1979).

Form and function of the macropodine stomach
Morphology

The external features of the stomach of *Macropus giganteus*, *M. eugenii* and *Thylogale thetis* are shown in Fig. 5.8. In contrast to the *M. giganteus* preparation in Fig. 5.1, here the mesentery has been dissected away from the stomach, and the organ has, in each case, been partially uncoiled for ease of comparison of the various gastric regions.

The sacciform region of the forestomach has been referred to by previous authors variously as the 'left cul-de-sac' (Owen, 1868), the 'left end of the stomach' or the 'cardiac fundus' (Schäfer & Williams, 1876), the 'blind sacs of the stomach' (Wilckens, 1872), the 'left cul-de-sac'

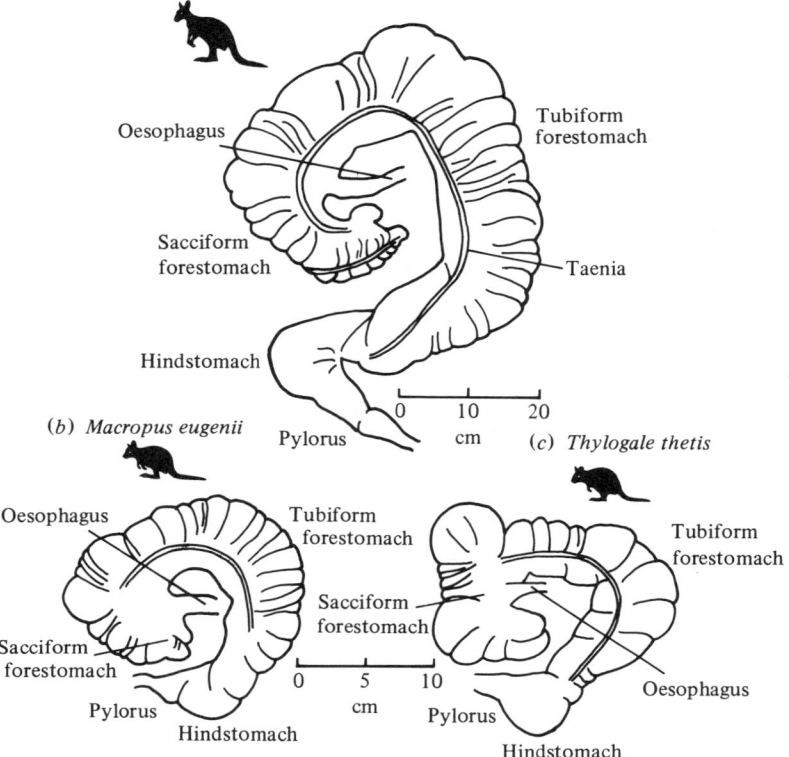

Fig. 5.8. The stomachs of *Macropus giganteus*, *M. eugenii* and *Thylogale thetis* cleared of mesentery and partially uncoiled to show external features. After Hume & Dellow (1980a).

or 'conical fundus' (Flower, 1872), and 'cul-de-sac' (Moir *et al.*, 1956). Hume (1978) and Kennedy & Hume (1978) referred to it simply as the 'forestomach'.

The tubiform region of the forestomach (the main tubular body of the organ) was called the 'midstomach' by Hume (1978) and Kennedy & Hume (1978). It corresponds to the 'middle stomach compartment' of Owen (1868), and the 'intestinal section of the stomach (der Darmtheil des Magens)' of Wilckens (1872). Other authors have not distinguished between the sacciform and tubiform regions as such. For instance, Griffiths & Barton's (1966) first gastric region of *Macropus rufus* includes the sacciform region with the cranial part of the tubiform region to the caudal end of the gastric sulcus (spiral groove). Their second gastric region consists of the rest of the tubiform region. Moir *et al.* (1956) included the sacciform region (cul-de-sac) of the stomach of *Setonix brachyurus* with the cranial section of the tubiform region, calling the combined region I the

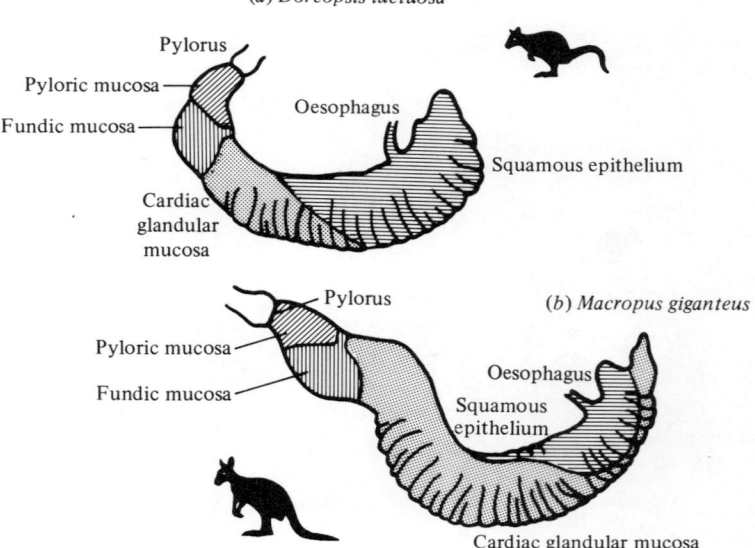

Fig. 5.9. The stomachs of *Dorcopsis luctuosa* and *Macropus giganteus* to show variations between species in distribution of the epithelial lining. Note that in *M. giganteus* the cardiac glandular mucosa of the sacciform forestomach is continuous with that of the tubiform forestomach. In *M. giganteus* dissected by Langer *et al.* (1980) the sacciform forestomach patch was separated from the tubiform forestomach cardiac mucosa by squamous epithelium. After Schäfer & Williams (1876).

forestomach, or 'rumen'. The caudal section of the tubiform region (region II) was termed the non-sacculated area.

Schäfer & Williams (1876), Oppel (1896) and Gemmell & Engelhardt (1977) divided the forestomach on the basis of histology rather than morphology. Thus region A of Schäfer & Williams (1876) and Oppel (1896) corresponds to the oesophaegeal region of Gemmell & Engelhardt (1977) in that it is lined by squamous epithelium. Region B (Schäfer & Williams, 1876; Oppel, 1896) is termed the cardiac region by Gemmel & Engelhardt (1977) because it is lined by cardiac glandular mucosa. The problem with this system is that the proportion of the forestomach wall lined with cardiac glands varies enormously among macropodine species, as can be seen in the drawings of the stomach of the New Guinea forest wallaby, *Dorcopsis luctuosa* (= *Dorcopsis muelleri luctuosa*) and of *Macropus giganteus* from Schäfer & Williams (1876) (Fig. 5.9).

Recently Richardson (1980) introduced a terminology as close as possible to that suggested in *Nomina Anatomica Veterinaria* of the International Committee on Veterinary Anatomical Nomenclature (1973). The stomach is divided into proximal, middle and distal compartments on the basis of the position of two of the three major gastric flexures, the caudal gastric flexure and the pyloric flexure (Fig. 5.10). Unfortunately the caudal gastric flexure is not always readily found in eviscerated specimens, and its functional significance is not clear. The proximal compartment includes the sacciform region of the *Macropus eugenii* forestomach; in *M. giganteus* it is likely to be more (see Fig. 5.8).

The hindstomach of Hume (1978), Kennedy & Hume (1978), Dellow (1979) and Langer *et al.* (1980) consists of the gastric pouch, the site of hydrochloric acid secretion, and the pylorus. It corresponds to the glandular pouch of Flower (1872), the pyloric fundus (region C) and that part of region B caudal to this (Schäfer & Williams, 1876), the fundic and pyloric regions of Gemmell & Engelhardt (1977), and the distal compartment of Richardson (1980).

In most macropodine species the tubiform forestomach is the most voluminous gastric region; for instance in *M. eugenii* it contains about 58% of the dry matter and in *M. giganteus* 72% of the total dry matter in the stomach (Langer *et al.*, 1980). However, in *T. thetis* and *T. stigmatica* (the Red-legged Pademelon) the most capacious stomach region is the sacciform forestomach as can be seen for *T. thetis* in Fig. 5.8. It constitutes 52% of total stomach capacity in *T. thetis* and probably a similar value in *T. stigmatica* (Langer, 1979). In contrast, *T. billardierii* (the Red-bellied Pademelon) lacks the dorsal pouch which is so prominent in the sacciform forestomach of the other *Thylogale* species above, and so this gastric region is not nearly so voluminous in *T. billardierii* (Dellow, 1979). Thus, even within the one genus, generalisations are often difficult to make.

In all macropodines the hindstomach is the smallest gastric region (Fig. 5.8).

The other major external features of the macropodine stomach are the taeniae and associated haustrations. The musculature of the forestomach (but not the hindstomach) wall is differentiated into three bands of longitudinal muscle, the taeniae, one on the left side, one on the right side, and the third under the line of attachment of the greater omentum on the greater curvature. Semi-lunar folds extend between the taeniae (Langer *et al.*, 1980), creating the haustrations which give the macropodine stomach its 'colon-like' appearance (Owen, 1868).

The semi-lunar folds are not permanent, but are formed by contractions of the stomach wall. Dellow (1979) described two main forms of contraction. The first are localised contractions involving each haustration and associated semi-lunar folds. They occur over a 4–6-second cycle and are

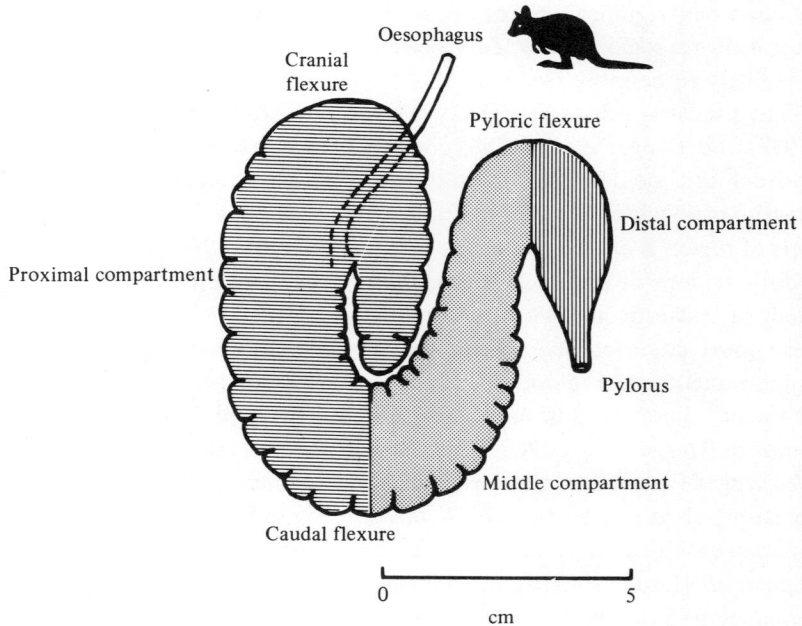

Fig. 5.10. Diagram of the stomach of *Macropus eugenii* to show the flexures on which Richardson's (1980) terminology of gastric divisions is based. After Richardson (1980).

independent but associated with sequential contractions of two or more adjoining haustrations. Their function appears to be local mixing of digesta. The second form of contraction is a stronger sequential wave of contraction that travels a short distance along the greater curvature wall. Each is a caudal displacement of a semi-lunar fold along the stomach wall, associated with relaxation of the successive caudal fold and formation of another fold cranial to the wave of contraction. This type of contraction seems to have more of a propulsive function.

Internally, there are differences among macropodine species in the presence and degree of development of the gastric sulcus, the position of the cardia (the opening of the oesophagus) in relation to the ventral fold between sacciform and tubiform forestomach regions, and the relative distributions of squamous and cardiac glandular epithelia.

The gastric sulcus

In ruminants an oesophageal groove (= reticular groove, ventricular sulcus) connects the cardia with the reticulo-omasal orifice. Contraction of its muscular walls creates a tube which allows milk in suckling young to bypass the reticulum and rumen where it could be fermented, and to pass directly into the omasum. Here it runs along the short omasal canal into the abomasum, the site of peptic digestion. It is indeed an ingenious device to allow for efficient utilisation of milk. The stimulus to closure of the sulcus is the suckling reflex. It does not appear to operate in adult ruminants unless they have been allowed to suckle throughout their development (Ørskov, Benzie & Kay, 1970).

In most macropodines a gastric sulcus extends caudally from the cardia along the lesser curvature of the tubiform forestomach. It is prominent in such species as *Macropus eugenii*, *M. robustus robustus*, *Wallabia bicolor* (the Swamp Wallaby) and *M. rufogriseus* (the Red-necked Wallaby) (Dellow, 1979). In *M. parma* and *M. rufus* only the right lip is well developed. In *M. giganteus* the sulcus is well developed in pouch young, but in adults it is relatively much smaller, and the lips poorly defined. This is an example of reduction during ontogenetic development.

No gastric sulcus can be found in *T. thetis* or *T. stigmatica*, in either adult or pouch young animals. However, it is present in *T. billardierii* (Dellow, 1979). The sulcus is also absent in *Peradorcus concinna* (the Nabarlek or Little Rock Wallaby), at least in adult animals; pouch young have not yet been examined (Hume & Dellow, 1980a).

It has been assumed that the gastric sulcus functions in suckling macropodine pouch young in a fashion similar to that described for suckling ruminants (Langer et al., 1980). However, evidence from Griffiths & Barton (1966) in the Red Kangaroo (shown in Plate 5.3) does not support this assumption. These workers found proteolytic activity to be distributed throughout the stomach of pouch young until 200 days of age. Only then was it restricted to the gastric pouch. The young leaves the pouch permanently at about 236 days (Sharman & Calaby, 1964). Histological observations supported the enzymatic findings. Columnar epithelium lined the interior of the whole stomach at birth. At 200 days of age there was a transition in histology between pouch young and adult. Gastric glands began to differentiate in the region destined to become the gastric pouch, and the gastric sulcus began to generate squamous epithelium. At 236 days of age true gastric tissue was found in the gastric pouch, and cranially the forestomach was lined with cardiac glands.

Thus it appears that the gastric sulcus, if it does function to channel ingested milk directly to the hindstomach, may be important in this regard only after the young has left the pouch for the last time. The now 'young at heel' red kangaroo will continue to suckle from the pouch for another 120 days (Tyndale-Biscoe, 1973).

Plate 5.3. Female *Macropus rufus* (Red Kangaroo) with large pouch young.

Recent research, however, has suggested a role for the gastric sulcus in adult macropodines. Dellow (1979) has been able to relate the pattern of initial distribution and subsequent dispersion of barium sulphate meals in the stomach of *M. eugenii*, *M. giganteus* and *T. thetis* to the relative sizes of the sacciform and tubiform regions of the forestomach, the presence or absence of a gastric sulcus, and the position of the cardia relative to the ventral fold between the sacciform and tubiform forestomach regions. His results, compiled from single radiographs and video-tape data, are shown in Fig. 5.11.

In *M. eugenii* the cardia opens on the caudal side of the sacciform–tubiform dividing fold. Consequently most of the contrast medium entering the stomach was directed into the tubiform forestomach region.

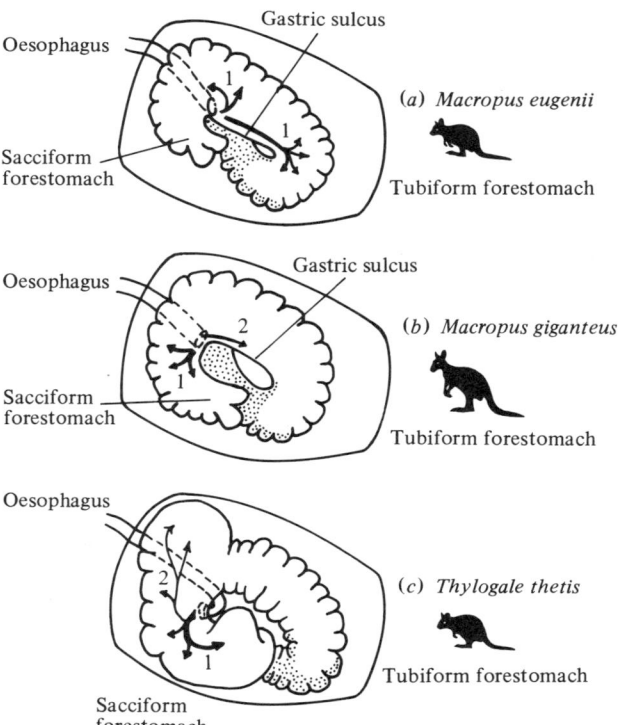

Fig. 5.11. Diagrams of the stomachs of *Macropus eugenii*, *M. giganteus* and *Thylogale thetis in situ* to show the dispersion of barium sulphate into the sacciform and tubiform regions, depending on the position of the cardia and the presence in *Macropus* of a gastric sulcus. 1, direction of primary dispersion; 2, direction of secondary dispersion. Stippled areas indicate residual barium sulphate 24 hours later. After Dellow (1979).

Here it moved caudally along the floor and immediate vicinity of the gastric sulcus, and mixed with digesta in the cranial and central regions of the tubiform forestomach. Contrast medium which entered the sacciform forestomach mixed with digesta close to the cardia.

In *M. giganteus* most of the barium sulphate entered the sacciform forestomach; in this species the cardia opens onto the sacciform–tubiform dividing fold. The contrast medium which did enter the tubiform forestomach was directed along the lesser curvature of the cranial region of this organ.

In *T. thetis* the cardia opens into the sacciform forestomach. Consequently all of the barium sulphate entered this gastric region, first mostly into the ventral area, later into the dorsal pouch. Up to an hour elapsed before any contrast medium appeared in the tubiform forestomach.

In all three species marked digesta was then gradually transported caudally along the tubiform forestomach, and eight hours after administration the whole of this region was outlined, and marked digesta was seen in the hindstomach. Food ingested six hours after barium sulphate administration did not mix with the contrast medium, reflecting the tubular nature of digesta flow, and the lack of one large single mixing pool in the kangaroo stomach.

Twenty-four hours after dosing, any contrast medium remaining in the stomach was confined almost entirely to the hindstomach. Contrast medium injected into the hindstomach outlined only this region of the stomach, and was transferred to the duodenum quite rapidly.

Thus it appears that the gastric sulcus probably plays a role in the nutrition of adult macropodines by assisting in the caudal movement of liquid digesta to the distal parts of the tubiform forestomach, a function first suggested by Owen (1868). This may be important in maintaining the fermentation rate in this part of the stomach.

We are now in a position to relate the faecal excretion patterns of [^{51}Cr]EDTA and [^{103}Ru]Phenanthroline in Fig. 5.5. to stomach structure in sheep and in the three macropodine species studied by Dellow (1979). The sacciform nature of the reticulo-rumen of sheep ensures that ingested food is effectively removed from the polarised flow of digesta along the digestive tract, and that it mixes with a large volume of digesta containing residues from previous meals. Differential flow of fluid and particulate digesta arises in part from a longer retention of larger food particles in the reticulo-rumen.

In contrast, although local mixing of digesta along the macropodine forestomach is effective, total mixing does not occur. Both the fluid and particles are subjected to tubular flow along the stomach, with extrusion of the fluid through the particulate digesta. Evidence suggests that all particles are transported along the stomach at the same rate (Dellow, 1979).

Of the excretion patterns shown in Fig. 5.5, the extremes are represented by the sheep, in which there is minimal separation of the two phases of digesta, and *M. giganteus*, in which this separation is greatest. This is probably due to: (*a*) the relatively small size of the sacciform forestomach region; and (*b*) the great absolute length of the tubiform forestomach compared with its diameter in the Eastern Grey Kangaroo. Together, these create the maximum tubular effect.

The excretion pattern of *T. thetis* is intermediate, and this can be explained in terms of the relatively greater size of the sacciform forestomach, which constitutes a large mixing pool, and a short tubiform forestomach.

The tubular nature of digesta flow through the macropodine stomach has important consequences for microbial digestion, as we shall see presently.

Gastric histology

Another aspect of the kangaroo stomach which shows considerable variation among species is the distribution of epithelial types. Schäfer & Williams (1876) first described the mucosa of the stomach of *M. giganteus* and *Dorcopsis muelleri luctuosa* and used this to delimit three gastric regions as discussed earlier (Fig. 5.9).

Squamous epithelium

The cellular structure of the squamous epithelium of the macropodine stomach is similar to that described for other mammals. Squamous epithelium from the sacciform forestomach of the Red-necked Pademelon and the Eastern Grey Kangaroo is shown in Plate 5.4. The basal cells are compact and cuboidal; they have few mitochondria and are separated by intercellular spaces. Towards the lumen of the stomach the cells are compacted and fibre-filled. The squamous epithelium of both species in Plate 5.4 is cornified. Gemmell & Engelhardt (1977) described the squamous epithelium from the Tammar Wallaby as being cornified also, but in neither their illustration nor that of Langer *et al.* (1980) is there any evidence of cornification in *M. eugenii*. The squamous epithelium interdigitates with the underlying lamina propria mucosae. Interdigitation appears to be most pronounced over permanent folds and in the gastric sulcus.

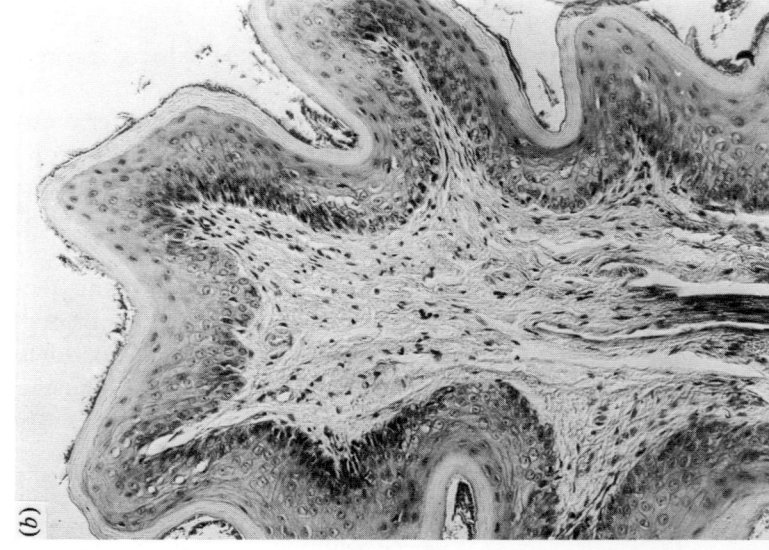
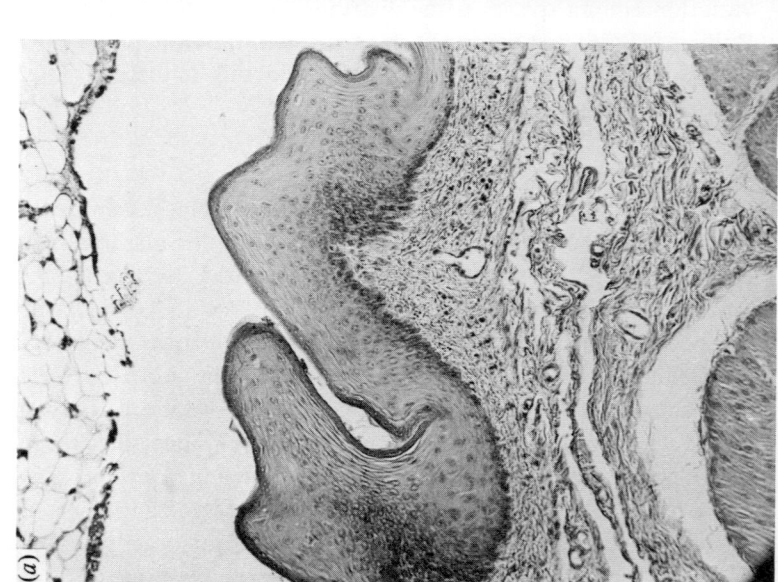

Plate 5.4. Squamous epithelium from the sacciform forestomach of (a) *Thylogale thetis* (×100) and (b) *Macropus giganteus* (×200). From Dellow (1979).

Cardiac glandular mucosa

The cardiac mucosa is several-fold thicker than the squamous epithelium. Sections of cardiac mucosa from *T. thetis* and *M. giganteus* (Plate 5.5) show the tubular mucin-secreting glands of this tissue. Griffiths & Barton (1966) found that water extracts of cardiac mucosa from the Red Kangaroo exhibited weak amylase activity, but considered that this was probably of microbial origin. No other enzymatic activity in adult cardiac mucosa has been reported.

Gastric glands

The epithelium of the gastric pouch is distinctly red in colour and strongly rugose. In section these folds are seen to be clothed with a deep epithelium of typical gastric tissue. Ultrastructural observations by Gemmell & Engelhardt (1977) confirm the presence of all four cell types normally associated with mammalian fundic mucosa; surface epithelial cells, mucous neck cells, chief cells (pepsinogenic cells) and parietal (hydrochloric acid secreting) cells.

Distribution of epithelial types

In the Tammar Wallaby squamous epithelium lines the floor and the lips of the gastric sulcus of the tubiform forestomach, and extends cranially to cover about 50% of the sacciform forestomach (Fig. 5.12). In the Eastern Grey Kangaroo tubiform forestomach only the floor of the gastric sulcus is lined with squamous epithelium, but, except for a small thick, isolated patch of cardiac glandular mucosa, all of the sacciform forestomach is covered with squamous epithelium. The area of squamous epithelium is more extensive in the Red-necked Pademelon, covering all of the sacciform region and the adjoining cranial border of the tubiform region. It extends further caudally in *T. stigmatica* (Langer, 1979) and *T. billardierii* (Dellow, 1979) to cover approximately 65% of the tubiform forestomach as well as the whole of the sacciform forestomach. *Dorcopsis muelleri luctuosa* is similar except that the border between squamous and cardiac epithelia is not perpendicular to the length of the tubiform forestomach (Schäfer & Williams, 1876) (see Fig. 5.9).

The illustration of *M. giganteus* of Schäfer & Williams (1876) (Fig. 5.9) indicates that the cardiac glandular mucosa extends in a narrow band along the greater curvature from the tubiform forestomach to the parietal blind sac of the sacciform forestomach. This differs from the observations of Owen (1868) and Langer *et al.* (1980), who remarked on the isolated nature

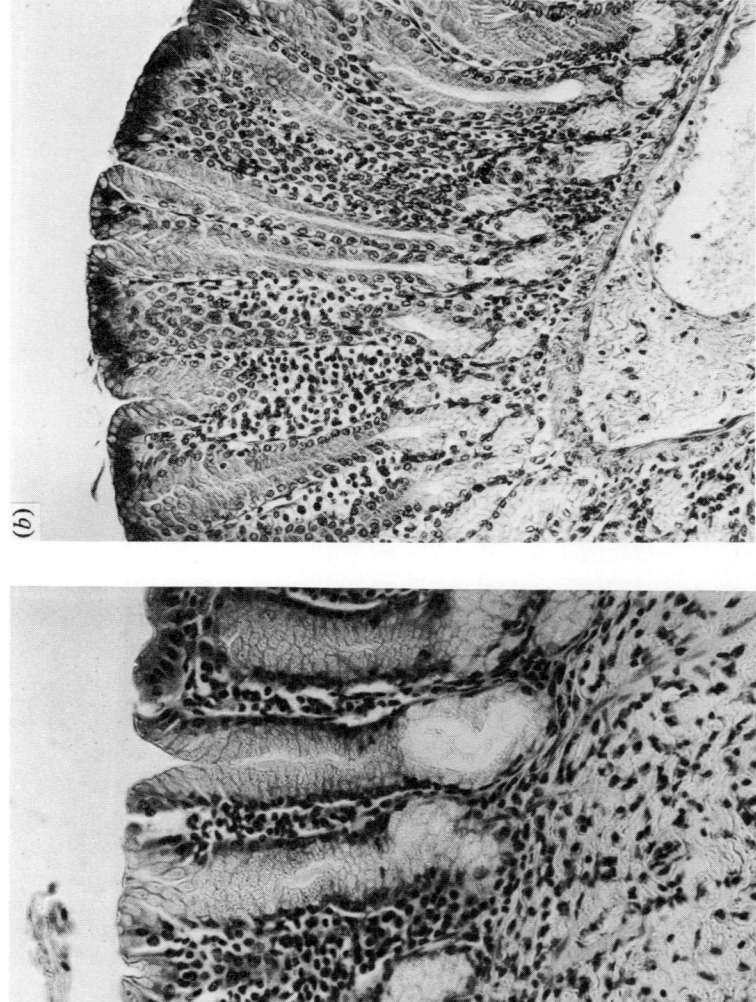

Plate 5.5. Cardiac glandular mucosa from the tubiform forestomach of (a) *Thylogale thetis* (×200) and (b) *Macropus giganteus* (×200). From Dellow (1979).

of the cardiac mucosa in the sacciform region. There is thus the possibility of minor intraspecific variations in the distribution of cardiac mucosa, and this is confirmed by observations by D. L. Obendorf (personal communication to Dellow, 1979) who has seen both forms in *M. giganteus* from Victoria.

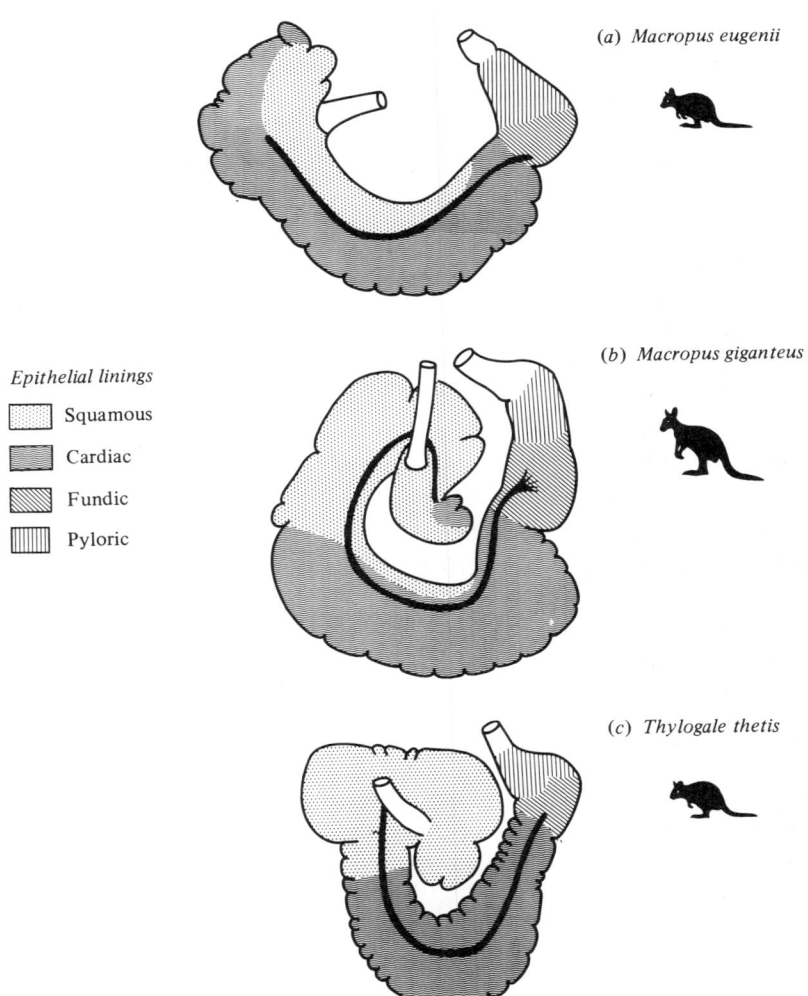

Fig. 5.12. Diagrams of the right-hand aspects of the stomachs of (*a*) *Macropus eugenii*, (*b*) *M. giganteus* and (*c*) *Thylogale thetis* to show the distribution of the four types of epithelial linings. After Langer *et al.* (1980).

The nutritional significance of these differences among species in the distribution of squamous and cardiac epithelia remains obscure. Further, whether the differences in extent of cornification of the squamous epithelium between species are independent of the nature of the diet is not known.

Fermentation and microbiology

Moir et al. (1956) provided the first preliminary description of the microbiology of the macropodine stomach in their study of the Quokka (*Setonix brachyurus*). They reported a 'dense bacterial population strikingly similar to that of the sheep's rumen under similar conditions. This population consisted mainly of Gram-negative rods and cocci, with a few spiral forms. Gram-positive rods were also present and these dominated the population where the pH was below 5.5.' Although only about 15 types of bacteria were discernible, compared with over 30 for the sheep, the total density of the population (10^{10} per ml) and the proportion of cellulolytic bacteria are similar (R. E. Hungate, personal communication to Moir, 1968). Dellow (1979) recorded direct counts of bacteria in forestomach digesta ranging from 5.7 to 76.0×10^{10} g^{-1} in Eastern Grey Kangaroos and 21.2 to 51.8×10^{10} g^{-1} in Red-necked Pademelons, all shot in the field. In Red-necked Pademelons and Tammar Wallabies maintained in captivity on chopped lucerne hay the direct counts ranged between 6.6 and 51.1×10^{10} g^{-1}; in sheep on the same diet the direct count from the rumen was 44.3×10^{10} g^{-1}.

Protozoa were not detected by Moir et al. (1956), but the authors remarked that this could possibly have been the result of the diet (commercial sheep pellets and chopped oaten hay). Subsequent work with grazing animals confirmed the presence of three unidentified ciliates at total concentrations of from 0.5 to 1.3×10^6 g^{-1} (Moir, 1965). Ciliates have also been found in the stomach of adult Red Kangaroos (Harrop & Barker, 1972) and in Tammar Wallabies (Lintern, 1970).

Dellow (1979) found ciliate protozoa in the forestomach of the majority of field and laboratory animals of six macropodine species. Total numbers were variable, ranging from 1.5 to 15×10^4 g^{-1}. The highest counts were always recorded in samples from the sacciform forestomach, and total numbers decreased progressively along the length of the tubiform forestomach, suggesting that either dilution rates were too high, or concentrations of soluble substrates decreased along the tubiform region. In some animals protozoa were found only in the sacciform forestomach and the most cranial region of the tubiform forestomach.

Moir (1965) remarked on the striking difference in the nature of the ciliate population between the Quokka and ruminants. In preliminary observations, Dellow (1979) thought that the majority of the macropodine ciliates structurally resembled the holotrichs *Dasytrica* and *Isotricha*. Heterotrichs, possibly two species, were also common. However, spirotrichs resembling *Entodinium* were found only in the sacciform forestomach of one *M. rufogriseus* and one *M. robustus robustus*.

Dellow (1979) also found fungal sporangia, similar to those found in ruminants (Bauchop, 1979a), in forestomach samples from all field animals except the Red-necked Pademelons; nor were they present in samples from *T. thetis* or *M. eugenii* fed chopped lucerne hay.

Interest in the ecology of fungi in the herbivore forestomach is very recent. Large numbers of anaerobic phycomycetous fungi are present in the reticulo-rumen of sheep and cattle fed roughage diets (Bauchop, 1979a, b). Sporangia are principally found attached to the more slowly digested fibrous material, and have not been found in digesta from ruminants grazing soft, leafy pasture. This may be used to suggest that the diet selected by *T. thetis* in the field is of consistently lower fibre content than the other macropodine species in Dellow's (1979) study. The highly selective feeding behaviour of *T. thetis* was noted by Johnson (1977) in his ecological studies on the species. However, factors other than fibre may also influence fungal colonisation in the forestomach, and more information on fungal ecology is needed.

Fermentation products
Volatile fatty acids

Moir *et al.* (1956) confirmed that the forestomach of the Quokka was a fermentation organ by reporting the occurrence of volatile fatty acids (VFA). The level was clearly related to time after feeding, varying from 22.5 mmol·l^{-1} in animals fasted for 22 hours to 105.1 mmol·l^{-1} in those recently fed. There was also a close relationship (Fig. 5.13) between VFA concentration and pH. The highest pH (8.0) in a fasted animal was close to the pH of saliva of 8.5.

The very low level of VFA in the hindstomach (3.2 to 10.6 mmol·l^{-1}) strongly suggested that the fermentation products were absorbed directly from the forestomach as in the ruminant. Absorption of VFA was supported by an increase after feeding of VFA levels in portal blood from 7.9 to 26.4 mg·dl^{-1}. Later work by J. M. Barker (1961) has confirmed the absorption of VFA from the Quokka stomach.

The proportions in which the individual VFA are found in the macropodine stomach appear to be quite similar to the situation in the rumen, with the possible exception that on the same diet the molar proportion of acetate tends to be lower, and that of propionate higher, in the macropodines (Table 5.4). The same holds true for the hindgut, the secondary fermentation area of both macropodines and sheep. It can be seen from Table 5.4 that the total VFA concentration is higher in the forestomach than in the hindgut, and that the molar proportion of acetate is lower, and that of propionate higher, in the forestomach than in the hindgut. This no doubt reflects the fact that substrate entering the hindgut is higher in structural carbohydrates and lower in readily fermentable carbohydrate than is the food ingested.

Similar molar proportions of VFA in the forestomach of the Red Kangaroo, as well as similar differences between forestomach and caecum, were reported by Henning & Hird (1970). However, Moir (1965) quoted much higher molar proportions (82–93%) of acetate and lower proportions of proprionate (2.2–6.4%) in the forestomach of grazing Quokkas. As Moir (1965) remarked, this is surprising in view of the facts that: (a) food intake appeared to be high (based on the finding that the average stomach contents were nearly 15% of body weight); (b) the early stage of growth of available pasture (ammonia levels in forestomach fluid averaged

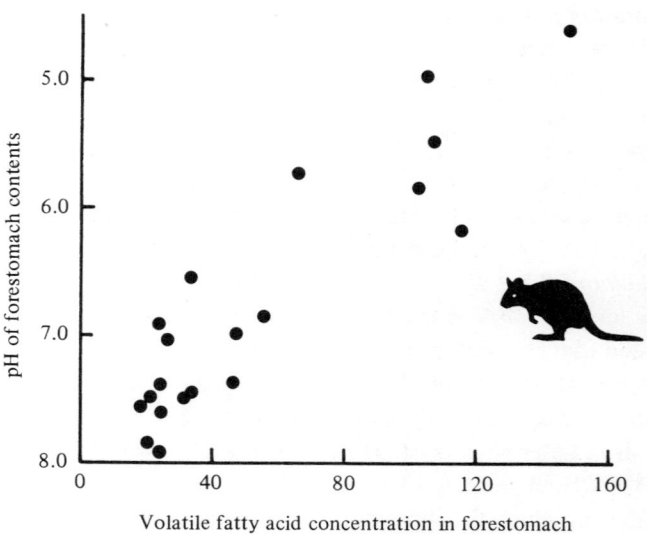

Fig. 5.13. The relationship between pH of forestomach digesta and concentration of volatile fatty acids in the forestomach of *Setonix brachyurus*. After Moir *et al.* (1956).

Volatile fatty acid concentration in forestomach contents (mmol · l^{-1})

33.4 mmol $N \cdot l^{-1}$); and (c) a high rate of fermentation (20.3 μmol $\cdot$ ml$^{-1} \cdot$ h^{-1}). All these factors favour a high proportion of propionate in the ruminant forestomach.

Dellow (1979) found that, in *T. thetis*, the concentration of total VFA was higher in the sacciform forestomach than in the cranial region of the tubiform forestomach. The fact that such a difference was not observed in *M. eugenii* or *M. giganteus* can be related to the relatively much larger mixing pool in the sacciform forestomach of *T. thetis* than in the other two species. In all three species the level of total VFA decreased along the length of the tubiform forestomach, in both laboratory and field animals. As suggested for the difference in VFA concentrations between forestomach and hindgut, the decrease along the tubiform forestomach reflects a progressive disappearance of the more readily fermentable components of the diet along the forestomach. This is illustrated in Fig. 5.14 in the Eastern Grey Kangaroo. Note particularly the very rapid disappearance of total soluble sugars; 95% was digested in the sacciform and cranial tubiform forestomach regions. Organic matter digestion was also more rapid in the

Table 5.4 *Concentration and molar proportions of volatile fatty acids in the forestomach and hindgut of wallabies and sheep fed chopped lucerne hay* ad libitum

	Thylogale thetis	*Macropus rufogriseus*	*Ovis aries*
Forestomach			
Total concentration (μmol $\cdot$ ml^{-1})	120	129	99
Molar proportion (% of total)			
Acetic	68	65	70
Propionic	19	21	18
Butyric[a]	10	9	9
Valeric[a]	3	5	3
Hindgut			
Total concentration (μmol $\cdot$ ml^{-1})	54	66	50
Molar proportion (% of total)			
Acetic	72	70	74
Propionic	15	16	17
Butyric[a]	10	10	6
Valeric[a]	3	4	3

After Hume (1977a).
[a] Includes both straight-chain and branched-chain isomers.

sacciform region, whereas acid-detergent fibre (ADF) was digested more slowly along the length of the forestomach. The initial rise in crude protein is indicative of recycling of endogenous nitrogen, probably mainly via saliva, to the forestomach, and incorporation into microbial protein.

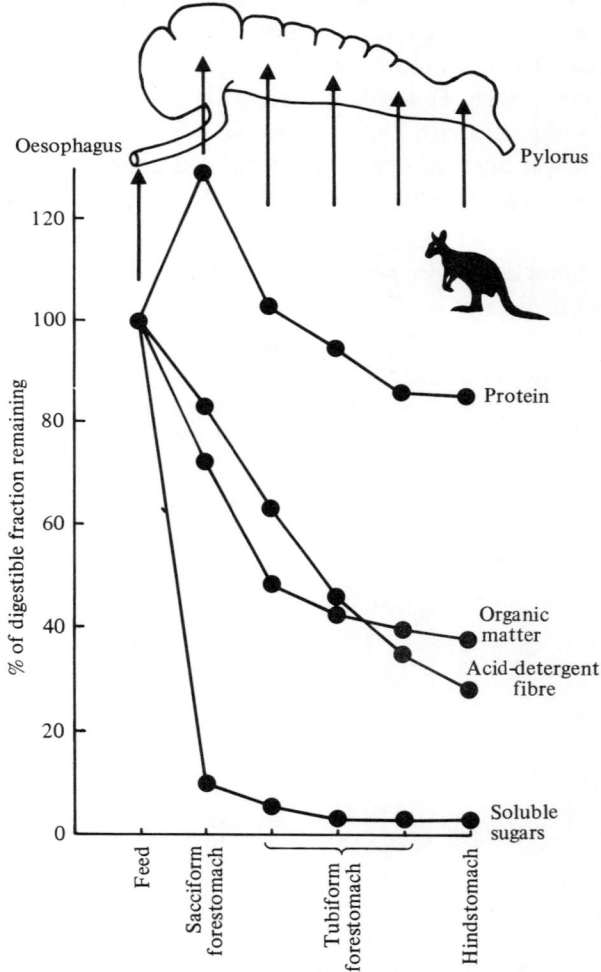

Fig. 5.14. The disappearance of digestible components of chopped lucerne hay along the stomach of *Macropus giganteus*. Note the very rapid disappearance of soluble sugars compared with fibre. The increase in crude protein in the sacciform forestomach is due to the incorporation of ammonia from endogenous sources into microbial protein. After Dellow (1979).

Ammonia

Ammonia is produced by many gastrointestinal bacteria (Prins, 1977), mainly from the deamination of dietary and microbial amino acids, and from urea. This ammonia can be the major source of nitrogen for bacterial protein synthesis in the sheep (Nolan & Leng, 1972).

Lintern-Moore (1973a) demonstrated the incorporation of from 64 to 85% of plant nitrogen into microbial protein in the forestomach of the Tammar Wallaby by following changes in the concentrations of various nitrogenous fractions in the forestomach with time after feeding. Her results are shown in Fig. 5.15. The decline in plant N and soluble N after feeding was quite rapid, and, concomitantly, there were increases in both bacterial and protozoal N. Bacteria constituted from 85 to 94% of total

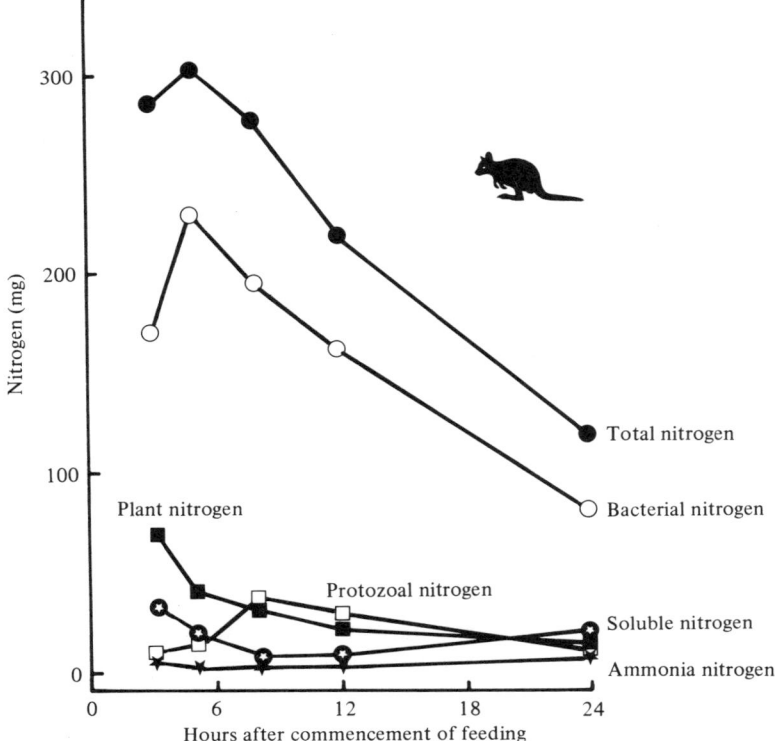

Fig. 5.15. Distribution of nitrogen in the forestomach digesta of *Macropus eugenii* at progressive intervals after commencement of feeding. After Lintern-Moore (1973a).

microbial N. Thus protozoa contributed less of the total microbial biomass than in the forestomach of the ruminant. This can be explained by the fact that protozoal concentrations fall along the length of the macropodine forestomach, and by the sampling procedure used by Lintern-Moore (1973a), in which digesta from the sacciform and tubiform forestomach regions were pooled for analysis.

There was no significant change in ammonia concentration in Lintern-Moore's (1973a) study. However, the findings of Brown (1969) and Lintern-Moore (1973b) that N from dietary urea was retained just as efficiently as N from casein, a readily degradable protein, in Euros and Tammars respectively, confirms that ammonia must be a key intermediate in microbial protein synthesis in the macropodine forestomach, just as it is in the rumen of the sheep.

Gases

Considerable quantities of gas are produced in the forestomach fermentation of macropodines. In the Quokka the gas consists of 65–75% CO_2, as well as hydrogen and methane (Moir, 1968). Kempton, Murray & Leng (1976) could not detect any methane in respired gas collected from Eastern Grey Kangaroos fed chopped lucerne hay, nor in the gas produced from *in vitro* incubation of stomach contents from other *M. giganteus* fed the same diet. However, Engelhardt, Wolter, Lawrenz & Hemsley (1978) found that Tammar Wallabies fed lucerne hay produced 6.5–11.0 ml methane per kg body weight per hour, equivalent to 1–2% of digestible energy intake. This is less than production in sheep, in which methane accounts for about 10% of digestible energy intake, but is still significant.

The only field data available are those of Dellow (1979). He found 5.4–9.5% methane in stomach gas collected from Red-necked Wallabies and Swamp Wallabies which were actively grazing or browsing when they were shot. Lower concentrations (0.5–1.8%) were found in Wallaroos and Eastern Grey Kangaroos, but these animals were resting. It was thus apparent that methane production was associated with feeding. Hydrogen concentrations were low (0.6–2.8%) in the Red-necked Wallabies, Wallaroos and Eastern Grey Kangaroos, but between 9.8 and 11.0% in the Swamp Wallabies. In all gas samples oxygen was always less than 0.1%, nitrogen less than 0.2%, and carbon dioxide greater than 70%.

The reason for the low or negligible methane production is not yet clear. Kempton *et al.* (1976) postulated that the absence of methane production in their Eastern Grey Kangaroos may have been due to oxygen entering the stomach along with food, or across the stomach wall, inhibiting the fermentation. Engelhardt et al. (1978) also referred to the possibility of significant diffusion of oxygen from the blood across the relatively thin stomach wall of the macropodines. An alternative, and more likely explanation, is that the relatively short retention time of particulate digesta and rapid flow of the fluid phase in the macropodine forestomach may prevent the establishment of the slow growing methanogens.

In ruminants methanogenesis serves as an efficient trap for excess hydrogen generated during microbial fermentation. With the possible exception of the Swamp Wallabies in Dellow's (1979) study, the low hydrogen concentration suggests that in the macropodine stomach hydrogen is either trapped in an alternative end product such as ethanol or formic acid, combines with oxygen diffusing across the forestomach wall, is absorbed across the stomach wall and eliminated in expired air, is transferred by passage of digesta to the intestine, or is removed by eructation. The definitive explanation awaits further research.

Rate of fermentation
In vitro *incubation studies*

Moir (1965) incubated Quokka stomach contents from six grazing animals for two hours and found an average increase in VFA of 20.3 μmol·g^{-1}·h^{-1}, a rate comparable with similar *in vitro* estimates in full-fed cattle (Carroll & Hungate, 1954). More recently Hume (1977a) measured VFA production *in vitro* in contents from the forestomach and the hindgut of two species of wallaby (*Thylogale thetis* and *Macropus rufogriseus*) and compared the results with sheep fed the same chopped lucerne hay diet. VFA production was in fact more rapid in the macropodine forestomach than in the ovine rumen (39 and 52 μmol·ml^{-1}·h^{-1} in the wallabies versus 23 in the sheep). VFA production was slower in the hindgut, but still faster in the wallabies (29 and 27 μmol·ml^{-1}·h^{-1}) than in the sheep (16 μmol·ml^{-1}·h^{-1}). As a proportion of the animal's digestible energy intake (i.e. energy absorbed), VFA production in the forestomach of the Red-necked Wallaby (42%) was greater than in either the Red-necked Pademelon (21%) or the sheep (29%). In the hindgut, because of the relatively greater size of the caecum and proximal colon in the sheep, total VFA production was greater in the sheep (6.9% of digestible energy intake) than in the wallabies (1.9 and 1.3%).

Presumably the faster fermentation rate in the macropodine stomach is a consequence of the higher turnover rate in that organ compared with the ovine rumen. This has been related to the very different structures of the macropodine and ruminant stomachs, as we have seen earlier, and, in the case of Hume's (1977a) study, to the smaller body size of the Red-necked Pademelon (5.5 kg) and Red-necked Wallabies (11.3 kg) compared with sheep (37.8 kg). Among ruminants Hungate et al. (1959) concluded that a fast fermentation rate, and a lower retention time, was necessary in small animals because of an apparent limitation to the size the reticulo-rumen could reach as a proportion of body size; in ruminants ranging in weight from 3.7 to 522 kg the contents of the reticulo-rumen averaged 12% of body weight.

In this regard it is interesting to note that VFA production *in vitro* in combined contents of the sacciform and tubiform forestomach regions of Eastern Grey Kangaroos (mean body weight 19.1 kg) was 29 μmol·ml^{-1}·h^{-1} (Hume & Dellow, 1980b), much slower than in the smaller species, and not much faster than in the sheep. Thus body size does seem to play a part in determining fermentation rate. However, this does not explain why VFA production was faster in *M. rufogriseus* than in the smaller *T. thetis*.

Much faster production rates have been observed in Red-necked Pademelons shot in the field while grazing on improved pasture adjacent to their rainforest refuge area. On this occasion sacciform forestomach and tubiform forestomach digesta were incubated separately. In two animals, VFA production was 102 and 99 μmol·ml^{-1}·h^{-1} in the sacciform region, and 60 and 65 μmol·ml^{-1}·h^{-1} in the tubiform region (Hume, unpublished observations). No doubt the higher concentration of rapidly fermentable carbohydrate in the fresh grass pasture compared with the chopped lucerne hay explains much of the difference in fermentation rates between laboratory and field. However, it is important to realise that there is always a significant time lag, up to 30 minutes (Hume, 1977a) between time of death and commencement of the incubation. In this period much of the soluble constituents of the lucerne hay are probably lost through fermentation, leading to an underestimate, during the two-hour incubation, of the actual VFA production rate. The effect would not be as important in the case of the fresh pasture material because of the much higher concentration of readily fermentable carbohydrate. The net result is probably an exaggeration of the difference in actual fermentation rates.

The limitations of *in vitro* estimations of fermentation rate in the forestomach were discussed in the previous chapter. There is no doubt that *in vitro* procedures underestimate VFA production, because of the time lag mentioned above, and because initial rates of fermentation immediately after ingestion of food are very high (Whitelaw *et al.* 1970). Thus Faichney

(1968) using an *in vitro* incubation technique, estimated that VFA production in the reticulo-rumen of sheep accounted for 34% of the digestible energy intake when fed a chopped lucerne hay diet. In contrast, estimates made *in vivo* by an isotope dilution technique in sheep fed similar diets yielded values of 54% (Gray, Weller, Pilgrim & Jones, 1967) and 53% (Leng, Corbett & Brett, 1968). Nevertheless, there is little reason to doubt that qualitative comparisons made between species or between different regions of the digestive system *in vitro* are valid, provided the animals are receiving the same diet.

Thus the markedly higher fermentation rate measured *in vitro* in the sacciform forestomach of wild *T. thetis* (page 148) compared with the tubiform forestomach is considered to be a real difference. Dellow (1979) found similar differences between these two forestomach regions in *T. thetis* and *M. eugenii* fed chopped lucerne hay in the laboratory, but again production rates were lower than in wild *T. thetis*.

Measurement of VFA production in vivo

Dellow (1979) successfully measured VFA production *in vivo* in the macropodine stomach, despite the problem of incomplete mixing of infused isotope due to the presence of more than one pool of digesta. The sacciform forestomach was considered to be a single pool of digesta. Since the isotope, [^{14}C]acetate, was infused into this pool this region was therefore the primary pool. The tubiform forestomach was considered as three successive pools in line with the sacciform forestomach, with no reversal of flow to a preceding pool. Flow of VFA along the stomach was estimated with reference to the fluid marker [^{51}Cr]EDTA. Total net production of acetate was then equal to the sum of net production, or irreversible loss (Leng, 1970) from all pools. From the molar proportions of the individual acids in the digesta, net production of total VFA was calculated.

The results are shown for *T. thetis* and *M. eugenii* in Fig. 5.16. In line with the *in vitro* results and the pattern of disappearance of organic matter (see Fig. 5.14), net VFA production and absorption decreased along the length of the forestomach. The ratio of VFA produced (mmol·day^{-1}) to intake of digestible organic matter (g·day^{-1}) was similar in both species (10.0 and 9.3 in *T. thetis* and *M. eugenii* respectively). Both are substantially higher than *in vitro* estimates (ratios of 2.4 and 2.7 mmol·g^{-1}), indicating again that the *in vitro* procedure underestimates true production. The *in vivo* results suggest that microbial fermentation in the macropodine stomach is as efficient as it is in ruminants (Czerkawski, 1978). The lower fibre digestion in macropodines must therefore be a function of shorter residence times in their forestomach.

Microbial protein synthesis

We have already seen that ammonia is a key intermediate in the synthesis of microbial protein in both the rumen and the macropodine forestomach. By infusing [^{15}N]ammonium chloride into the sacciform forestomach along with the [^{14}C]acetate and [^{51}Cr]EDTA, Dellow (1979) estimated net microbial protein synthesis in both the Red-necked Pademelon and Tammar Wallaby concurrently with his measurements of VFA

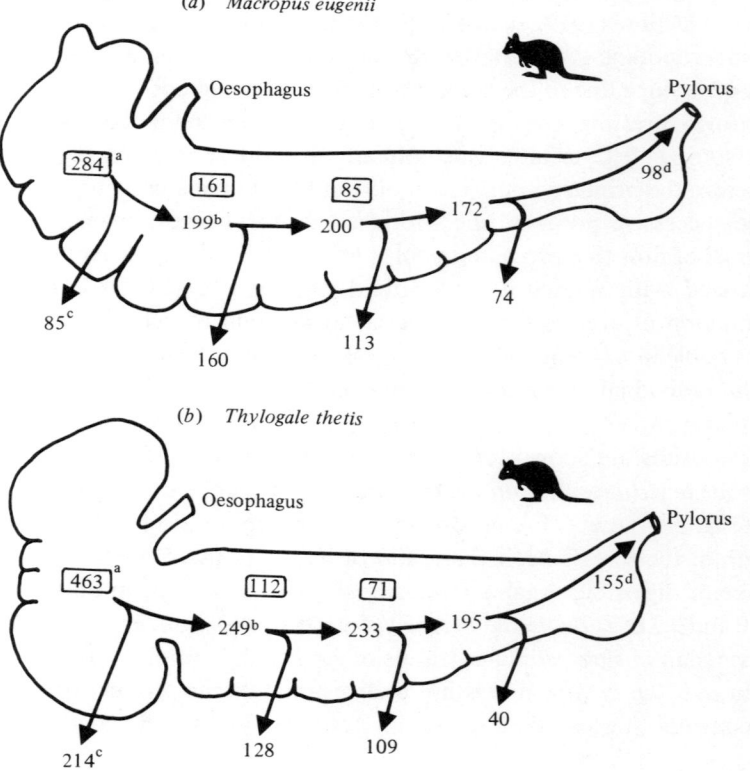

Fig. 5.16. Diagram showing the production, absorption and flow of volatile fatty acids (mmol·day^{-1}) in the stomach of *Macropus eugenii* and *Thylogale thetis*.[a] (in boxes) net production; [b] flow along the forestomach; [c] net absorption; [d] flow through the hindstomach. From the *in vivo* experiments of Dellow (1979).

production. After 48 hours of infusion, considered to be more than sufficient for steady-state labelling of microbial protein with ^{15}N, the animals were slaughtered and the flow of bacterial nitrogen arising from ammonia that passed through the hindstomach was calculated from the relationship:

$$\text{Bacterial flow (gN·day}^{-1}) = \text{non-ammonia N flow (g·day}^{-1}) \times \frac{\text{non-ammonia enrichment in HS}}{\text{bacterial enrichment in TFS}}$$

where HS = hindstomach and TFS = tubiform forestomach.

This procedure assumes that enrichment of bacterial nitrogen with ^{15}N does not change significantly between the central region of the tubiform forestomach and the hindstomach.

In the sacciform forestomach the proportion of bacterial nitrogen derived from ammonia was 44% in *T. thetis* and 40% in *M. eugenii*; in the tubiform forestomach this proportion increased to 74% and 84% in the two species respectively. The lower values in the sacciform forestomach are indicative of more direct incorporation of peptides and/or amino acids of either dietary or endogenous origin, or both.

Dellow's (1979) estimates of net microbial protein synthesis (25 g nitrogen per kg organic matter apparently fermented in the stomach of *T. thetis* and 27 in the stomach of *M. eugenii*) are within the range of values from ruminants (Czerkawski, 1978). Thus, as we have seen with VFA production, microbial protein production in the macropodine stomach is at least as efficient as in the reticulo-rumen of ruminants.

VFA and carbohydrate metabolism

In his studies Dellow (1979) found that between 24 and 41% of acetic acid formed in the macropodine stomach was converted to butyric acid. This is similar to estimates in the sheep (Leng & Leonard, 1965).

The metabolism of VFA by the rumen wall has been investigated in some detail (Annison & Armstrong, 1970). It has been shown that, under aerobic conditions, rumen epithelium metabolises butyrate to a much greater extent than either acetate or propionate. Most of the butyrate is converted into ketone bodies, mainly acetoacetate. Henning & Hird (1970) showed that metabolism of butyrate in the mucosa of the forestomach of the Red Kangaroo and the Eastern Grey Kangaroo is analogous to that in the rumen wall, and in the guinea pig caecum. From this they concluded that the epithelium of all fermentative organs probably converts butyrate

152   5 *Digestion and metabolism in kangaroos*

produced in the organ into ketone bodies. The partial oxidation of butyrate could satisfy the energy needs of the epithelial tissue and at the same time provide a substrate, ketone bodies, to other tissues for further oxidation. Another consequence of ketogenesis is the regeneration of coenzyme A, which could otherwise limit the rate of oxidation in cells heavily loaded with fatty acids (Henning & Hird, 1970). The only odd result in Henning & Hird's (1970) study was the suggestion that in the macropodine stomach the squamous epithelium showed negligible ketogenic activity, while the cardiac glandular mucosa had uniform activity. This contrasts with the reticulo-rumen, which is lined entirely by squamous epithelium, but which exhibits ample ketogenic activity throughout.

Several findings indicate that the carbohydrate metabolism of adult macropodines is similar in many ways to the ruminant system. Kerry (1969) measured the activity of various disaccharidases in homogenates of small intestinal mucosa from a range of marsupials. The levels of activity

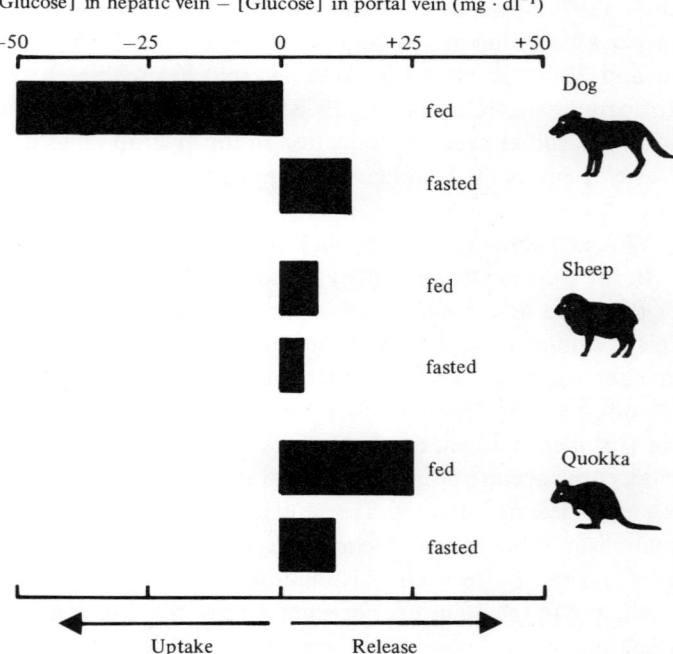

Fig. 5.17. Differences between glucose concentration in the hepatic vein and the portal vein in fed and fasted dogs, *Setonix brachyurus* and sheep. Both the Quokka and the sheep release glucose into the blood from the liver at all times. After Ballard *et al.* (1969).

are shown in Table 3.1. Clearly the low levels in the Eastern Grey Kangaroo indicate that little digestible carbohydrate normally reaches the small intestine. This agrees with the rapid disappearance of total soluble sugars along the forestomach in Fig. 5.14. It also complements the observations of Moir et al. (1956) that, like ruminants, Quokkas have lower blood glucose levels than simple-stomached animals, of Ballard, Hanson & Kronfeld (1969) that the rate of glucose incorporation into glycogen in the liver of the Quokka is less than 4% of the rate found in rat liver (Table 5.5) and of J. M. Barker (1961) that the Quokka apparently releases glucose into the blood at all times, as does the sheep; this is in contrast to the fed dog, in which glucose is taken up by the liver. This difference is shown in Fig. 5.17 (Ballard et al., 1969).

The finding that the livers of sheep and the Quokka do not take up glucose from the blood is further reflected in the absence of glucokinase activity (Table 5.5). Measurement of the activities of the two glucose phosphorylating enzymes in rat liver have shown that hexokinase is saturated with respect to substrate at low glucose concentration, while glucokinase has a maximum activity at much higher glucose concentrations. During long-term fasting, glucokinase activity is lost while hexokinase activity remains unchanged (Ballard et al., 1969). Since glucokinase is both the adaptive and the more active glucose phosphotransferase in rat liver, ruminants and macropodines have apparently adapted to the low glucose

Table 5.5. *Hexokinase activity, glucokinase activity and glucose incorporation into glycogen in liver slices*

	Hexokinase units[a] $\cdot$ g^{-1}	Glucokinase units $\cdot$ g^{-1}	Glucose incorporation μmol $\cdot$ g^{-1} $\cdot$ h^{-1}
Omnivores			
Rat	0.66	3.83	25.4
Mouse	0.70	3.08	7.4
Pig	0.64	3.17	5.8
Herbivores – hindgut fermenters			
Guinea pig	0.39	1.25	6.5
Rabbit	0.28	1.32	11.8
Herbivores – forestomach fermenters			
Sheep	0.13	< 0.03	0.3
Cow	0.39	< 0.03	0.2
Quokka (*Setonix brachyurus*)	0.39	< 0.03	1.0
Tammar (*Macropus eugenii*)	0.21	< 0.03	0.6

After Ballard et al. (1969).
[a] One unit of enzyme activity hydrolyses 1 μmol substrate $\cdot$ min^{-1}.

absorption from the gut by a loss of a major portion of the glucose phosphorylating activity.

Like the sheep, the Quokka (J. M. Barker, 1961) and Red Kangaroo (Griffiths, McIntosh & Leckie, 1969) show considerable tolerance to the effects of large intravenous insulin injections, even though their blood glucose levels are substantially reduced. Conversely, Red Kangaroos made diabetic by destroying the Islets of Langerhans in the endocrine pancreas with injections of alloxan, develop a marked hyperglycaemia (300 mg·dl^{-1}) and die unless treated with insulin (Griffiths et al., 1969). This contrasts with the much greater tolerance of simple-stomached animals, both eutherian and marsupial, to hyperglycaemic conditions (Table 5.6).

That propionate is a gluconeogenic precursor in macropodines was demonstrated by J. M. Barker (1961). She found that there was no increment in blood glucose level following intravenous acetate injections, but a noticeable rise when propionate was administered. This is similar to the responses of the sheep and the rabbit, and reflects the marked glucogenicity of propionate.

Blood glucose levels in the Red Kangaroo were reported by Griffiths et al. (1969) to be unresponsive to daily injection of cortisone for six days; rabbits and sheep responded in a manner similar to that shown in other eutherians, with an increase in blood glucose concentration. The findings in the Red Kangaroos were thus of considerable interest, suggesting that there may be a fundamental endocrine difference between eutherians and marsupials. However, since then a classical eutherian-like response has been reported to cortisol in the Brushtail Possum (*Trichosurus vulpecula*) (Than & McDonald, 1973) and to cortisol and to adrenocorticotrophin

Table 5.6. *Responses to changes in blood glucose levels in marsupial and eutherian herbivores*

	Plasma glucose (mg·dl^{-1})	Response to	
		Hypoglycaemia (Insulin)	Hyperglycaemia (Alloxan)
Forestomach fermenters			
Setonix brachyurus	78	Coma at 20	—
Macropus rufus	73–85	No effect at 50	Coma at 300
Ovis aries	65	No effect	Coma
Hindgut fermenters			
Trichosurus vulpecula	120	Coma at 40	No effect below 400
Oryctolagus cuniculus	120	Coma at 50	No effect

After Tyndale-Biscoe (1973).

(ACTH) in the Tammar Wallaby (Cooley & Janssens, 1977); cortisol is the major glucocorticoid produced by both Brushtail (Weiss & McDonald, 1966) and Tammar adrenals (Catling & Vinson, 1976). Thus the Red Kangaroo remains the only mammal in which glucocorticoids have been reported to be ineffective in raising blood glucose levels.

Lipid metabolism

The presence of *trans* acids in the depot fat of Red and Eastern Grey Kangaroos (Hartman, Shorland & McDonald, 1955) supports the view that microbial modification of dietary constituents in the macropodine forestomach is extensive. In ruminants depot fats are highly saturated, the degree of saturation being altered significantly only if unsaturated dietary lipid is protected from microbial hydrogenation in the reticulo-rumen in some way (Cook, Scott, Ferguson & McDonald, 1970). In the kangaroos there was a considerable di- and poly-unsaturated fatty acid content in the depot fat, indicating that although microbial modification of dietary fats is substantial, it is less than in ruminants. This is consistent with the shorter retention times of dietary constituents in the macropodine stomach compared with the reticulo-rumen (see page 124). Redgrave & Vickery (1973) found that the fat in meat from horses and Eastern Grey Kangaroos both had a ratio of polyunsaturated to saturated fatty acids of about 1:1, which is considerably higher than the ratios reported for cattle fed protected polyunsaturated fat diets (Cook *et al.*, 1970).

Nitrogen metabolism and urea recycling

Ruminants conserve nitrogen when dietary supplies are low by utilising endogenous urea and protein as sources of nitrogen for microbial protein synthesis (Houpt, 1970). Studies using ^{15}N have revealed that ammonia production in the rumen may be equivalent to 17–84% of dietary nitrogen intake (Mercer & Annison, 1976). Ammonia serves as a source of nitrogen for bacterial growth in the rumen to the extent that 50–80% of microbial nitrogen may be derived from ammonia. Kennedy & Hume (1978) estimated that a similar proportion (63%) of bacterial nitrogen in the tubiform forestomach of the Tammar Wallaby was derived from ammonia.

Lintern (1970) followed the fate of intravenously injected [^{15}N]urea in Tammar Wallabies, particularly the relationship between renal retention and recycling to the gut. She gave single injections of [^{15}N]urea to animals fed either a high (2.60%) nitrogen or a low (0.34%) nitrogen diet *ad libitum*. The injection was given 8 hours after feed had been removed, and the animals were anaesthetised throughout the three-hour collection period. As can be seen from Table 5.7, much less of the injected dose of ^{15}N was excreted in the urine, and more was detected at slaughter in the stomach and caecum of animals fed the low protein diet. Utilisation of urea-nitrogen recycled to the stomach, as measured by incorporation into the bacterial fraction, was also higher in the Tammars fed the low protein diet. In agreement with other estimates of the importance of protozoa in the macropodine stomach (Dellow, 1979), incorporation of ^{15}N into the Tammar stomach protozoal fraction was insignificant.

Kennedy & Hume (1978), using intravenous infusions of urea labelled with either ^{15}N or ^{14}C, also showed that urea was transferred from the blood to the gut in Tammar Wallabies. The proportion of urea synthesised in the liver which was transferred to the gut was similar and high (74–86%) on both a high nitrogen chopped lucerne hay and a low nitrogen chopped oaten hay diet. However, incorporation of nitrogen from endogenous urea into microbial protein in the gut was equivalent to only 34–53% of dietary nitrogen intake on the high nitrogen diet, while on the low nitrogen diet incorporation was equivalent to 103–112% of nitrogen intake. This latter result indicates that urea recycling in macropodines may be of sufficient magnitude to sustain microbial function in the gut during periods of nitrogen deprivation, in the same way as has been demonstrated in ruminants (Kennedy & Milligan, 1980).

That the macropodine kidney concentrates urine and retains urea on low nitrogen diets was simply demonstrated by Lintern & Barker (1969). They fed two groups of Tammars either a high (1.2% nitrogen) or low (0.4%

Table 5.7. *Fate of [^{15}N]urea injected intravenously into Tammar Wallabies fed either low or high nitrogen diets*

Animal	Diet	Urea (mg) injected	Injected urea (mg) appearing after three hours in		
			Urine	Fore-stomach	Caecum
1	Low	463	2	62	1.5
2	nitrogen	471	2	63	1.9
3	(0.34%)	604	2	59	1.6
4	High	471	214	16	0.6
5	nitrogen (2.60%)	532	245	19	0.6

After Lintern (1970).

nitrogen) protein diet for 28 days. In the high nitrogen group, which remained in postive nitrogen balance throughout, plasma urea concentration remained unchanged, and the ratio of concentration of urea-nitrogen to total nitrogen in the urine, the UR ratio (Kinnear & Main, 1975), was consistently 0.85. The low nitrogen group dropped into negative nitrogen balance, plasma urea concentrations fell to half their initial value, and the UR ratio fell to about 0.35, indicating that proportionately less urinary nitrogen was in the form of urea, which presumably was being retained by the kidneys.

At the end of the 28-day collection period Lintern & Barker (1969) slaughtered the animals and sliced the kidneys while frozen. A central slice was cut into six sections, each section was then weighed and ground in a mortar with distilled water, and urea concentrations in each homogenate assayed. Their results are shown in Fig. 5.18. The concentration of urea

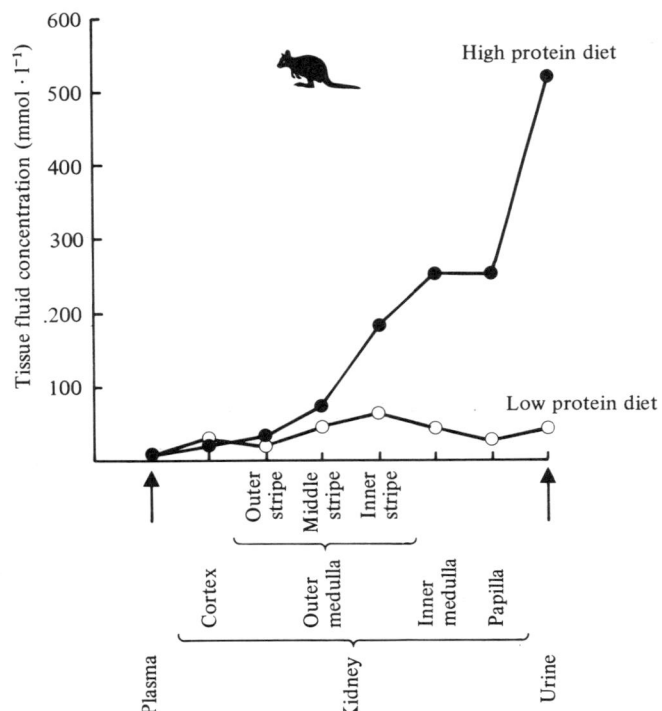

Fig. 5.18. The concentration of urea in the plasma, kidney and urine of *Macropus eugenii* fed diets of high or low protein content. The low urine:plasma ratio on the low protein diet suggests that urea is being retained by the kidney as shown in the diagram. From data of Lintern & Barker (1969).

in renal tissue water of the high nitrogen group increased from cortex to papilla but was lower than in urine. In the low nitrogen group however, a peak concentration of urea was reached in the inner stripe, outer zone of the medulla and in two of the three animals was higher there than in urine. A similar pattern has been found in sheep on low nitrogen diets (Schmidt-Nielsen & O'Dell, 1959), indicating that Tammar Wallabies retain urea in their kidneys in a manner similar to sheep and camels (Schmidt-Nielsen, 1958). In the rat it has also been found that differences in the concentration of urea in renal tissue occur between animals fed high and low nitrogen diets, but on the low nitrogen diets the highest concentration of tissue urea is found in the inner medulla, indicating poor ability to retain urea under nitrogen stress (Lintern & Barker, 1969).

Conclusion

The outstanding feature of the digestive physiology of macropodines is that they are all foregut fermenters, so that food is subjected to extensive microbial action before being exposed to the animal's own enzymes. As in other foregut fermenters this enables the macropodines to utilise structural carbohydrates as a food source, something which animals without a sizeable microbial population somewhere in their gut cannot do.

Despite earlier descriptions of the digestive physiology of macropodines as 'ruminant-like', there are numerous features of the macropodine gut which are very different from those of the Ruminantia. Their gastric morphology leads to tubular flow of digesta through the stomach and to shorter passage times of food residues through the gut. Estimates of fermentation rate and microbial protein production suggest that microbial fermentation in the macropodine forestomach is at least as efficient as it is in ruminants. The lower fibre digestion in macropodines must therefore be a function of shorter residence times in their gut, primarily in the forestomach.

In other aspects of their metabolism macropodines appear similar to ruminants. That is, their carbohydrate metabolism is adapted to low glucose absorption from the gut and propionic acid has been shown to be a gluconeogenic precursor. Nitrogen metabolism in macropodines also appears to be similar to that in ruminants in that urea is retained by the kidneys on low protein diets. Recycling of urea to the gut, and its incorporation into microbial protein probably plays a significant role in sustaining microbial function in the macropodine gut during periods of nitrogen deprivation.

6

Diet and nutrition of kangaroos and wallabies

This chapter provides an opportunity to discuss some aspects of the ecology of macropodines in relation to the knowledge we now have of some of their nutrient requirements, principally energy, protein and water (Chapter 1) and of their digestive and metabolic adaptations to herbivorous diets (Chapter 5). However, before considering specific examples of field nutritional studies, some emphasis should be placed on a discussion of food preferences and feeding behaviour in relation to the major habitat types in which the macropodines are found.

Dentition and diet (grazer versus browser)

The relationship between dentition and diet was first examined by Raven & Gregory (1946) in their discussion of the possible evolution of the Macropodidae. This relationship has recently been re-examined by Sanson (1976, 1978) who classified the macropodines, both extant and extinct, into three grades, viz. browsing, intermediate and grazing, on the basis of dental morphology. The basic assumption is that tooth morphology reflects dietary adaptations. Examination of all genera of the Macropodinae except *Dorcopsoides* led Sanson (1978) to believe that there were two basic types of masticatory organisation, representing an ancestral browsing grade and a derived grazing grade. Browse was defined as soft, unabrasive, low fibre herbage, while graze consists of abrasive, siliceous grasses often of high fibre content.

The two basic types of dental morphology are shown in Fig. 6.1, first showing the initial contact between opposing lophs, and again at maximum interdigitation. Comparison of the two types shows that the weak longitudinal ridges (links) connecting the transverse lophs of type-B result in a crushing action; there is a large surface area of contact between the upper and lower teeth. This is characteristic of browsing species (Sanson, 1980).

In contrast, Type-G molariform teeth are characterised by strong links between the lophs. When occluded, the well-developed links are in contact with opposing lophs, such that the surface area of contact is reduced, resulting in more of a cutting action, ideal for comminuting higher fibre material. This is characteristic of grazing species (Sanson, 1980).

The two grades established by Sanson (1978) are shown in Table 6.1. Since Sanson (1978) considers that the grazing grade evolved from the browsing grade, recognition of some intermediate types is to be expected. Thus, among living forms, *Petrogale* may be a browsing form evolving into a grazing type.

How does Sanson's (1978) classification agree with what we know about the feeding habits of macropodine groups? In many instances this

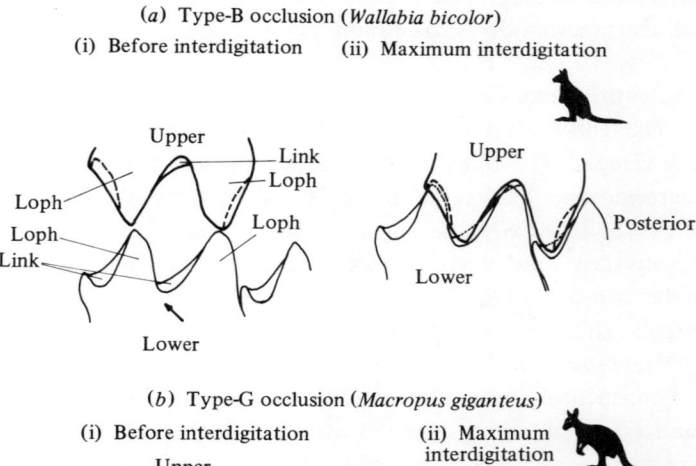

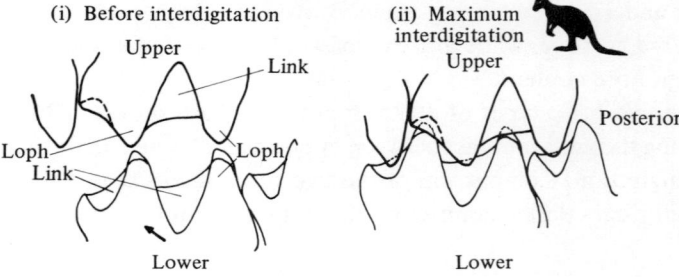

Fig. 6.1. Molariform teeth from *Wallabia bicolor*, a browser, and *Macropus giganteus*, a grazer, showing the initial contact between opposing lophs and maximal interdigitation. The arrows indicate the direction of movement of the lower jaw. Note the large surface area of contact in *W. bicolor* compared with *M. giganteus*. After Sanson (1980).

is extremely difficult to test, because of the changes in the Australian landscape and vegetation cover wrought by European man. Thus the habitat in which *Thylogale thetis* is now found is nutritionally quite different from that in which it evolved. However, in other cases we have a good idea of natural feeding habits. The most comprehensive dietary studies have been concerned with the large kangaroos *Macropus giganteus*, *M. rufus* and *M. robustus erubescens*, particularly in relation to the extent of possible competition between kangaroos and domestic stock.

The techniques used

The problems associated with determining the diets of carnivores and omnivores have already been mentioned in Chapters 2 and 3 respectively. In herbivores, especially forestomach fermenters, the problem is compounded by differential rates of fermentation of structural versus non-structural components of plant material. In studies of macropodid diets a variety of techniques for identifying what plant species are taken, and in what proportions, has been used. Storr's (1961) original method was based on digesting faeces in acid, sampling the remaining particles, and identifying and measuring the surface area of leaf epidermis microscopically. The main problem with this method is the differential loss of species during digestion in the animal and by acid-treatment of the faeces. Annual plant species did not survive passage through the gut of the Quokka (*Setonix brachyurus*) very well at all (Storr, 1964a). Small sample

Table 6.1. *Classification of macropodine genera on the basis of dental morphology*

Browsing grade	Intermediate	Grazing grade
Wallabia	Petrogale	Macropus
Dorcopsis	Tropsodon[a]	Peradorcus
Dorcopsulus		Onychogalea
Dendrolagus		Lagorchestes
Thylogale		Fissuridon[a]
Setonix		Procoptodon[a]
Lagostrophus		
Hadronomus[a]		
Dorcopsoides[a]		
Prionotemnus[a]		
Sthenurus[a]		

After Sanson (1978).
[a] Extinct genera.

162   6 *Diet and nutrition of kangaroos and wallabies*

sizes, only two microscope slides from one faecal pellet per animal, have also been cited as criticism of Storr's (1961) technique. Nevertheless, it was used by Ealey & Main (1967) in their study of seasonal changes in the nutrition of the Euro in north-western Australia with apparent success. This is possibly because of the predominance of only a relatively few species in the available pasture and thus in the diet.

Griffiths & Barker (1966), working on Red and Eastern Grey Kangaroos in a more diverse botanical habitat in south-western Queensland, were dissatisfied with Storr's (1961) technique, and used instead microscopic examination of stomach contents from animals shot while feeding. Although samples of stomach contents were dried and ground, acid digestion was not used, and so no parts were destroyed chemically, although the possibility still remained that readily fermentable species and plant parts may not have been recovered quantitatively. This was recognized by Bailey, Martensz & Barker (1971), who used Griffiths & Barker's (1966) technique in combination with samples of mouth contents of Red Kangaroos shot in north-western New South Wales. Griffiths, Barker & Maclean (1974) went back to a technique based on faecal samples, since macropodid numbers during a second study were very low, partly as a result of a severe drought. After identifying plant particles in faeces of

Plate 6.1. The rare Yellow-footed Rock-wallaby (*Petrogale xanthopus*) basking in the sun. (Ray Williams.)

animals fed a diet of known composition, and comparing the proportions of different plants in the stomach and rectal faeces of shot animals, these workers concluded that the faecal method was suitable for comparison of the relative proportions of plants eaten by different species, assuming that their digestive processes are similar, but not for determination of the absolute proportions of different plants in the diet of a single species. Dawson & Ellis (1979) reached a similar conclusion before using faecal analysis to ascertain the diet of the rare Yellow-footed Rock-wallaby (*Petrogale xanthopus*) (Plate 6.1) in western New South Wales when shooting of animals to obtain gut samples was out of the question.

The reliability of these methods can probably be explained by the robust nature of the plants found in arid and semi-arid environments, in which all the work described above has been done. With softer diets the techniques may not be valid at all (Slater & Jones, 1971).

The techniques used rely upon diagnostic characteristics of epidermal material, the most resistant parts of arid-zone plants. Griffiths & Barker (1966) and Griffiths, Barker & Maclean (1974) used cuticle pattern, hairs, asters, silica cells and spines after washing material in alcohol and xylol for plant species identification. In their study of seasonal changes in diet preferences of Red Kangaroos, Euros and sheep in western New South Wales, Ellis, Russell, Dawson & Harrop (1977) modified this procedure by incorporating a preliminary bleaching and staining procedure which greatly facilitated identification. Although not all particles containing epidermal fragments could be identified to plant species, generally they could be easily separated into six broad groups, viz.:

(1) Grasses, distinguished by regular, parallel cell patterns.
(2) Flat-leaved chenopods, characterised by vesicular hairs, and consisting basically of saltbushes (*Atriplex* and *Rhagodia* spp.).
(3) Round-leaved chenopods, mostly distinguished by simple uniseriate hairs and largely comprising bluebushes and copperburrs (*Maireana* and *Bassia* spp.).
(4) Browse, trees and large shrubs which are identified by clearly defined small cells, often heavily lignified, and by recognisable cell arrangements and accessories.
(5) Plants with stellate trichomes, distinguished by multi-armed hairs, mostly star-shaped, and consisting of members of the family Malvaceae, *Solanum* spp. and *Ptilotus obovatus* (mainly small shrubs).
(6) Forbs, non-woody plants other than grass, including many small ephemeral species such as composites and papillionates.

Nevertheless, in Ellis *et al.*'s (1977) study between 28 and 68% of total particles from the stomach could not be identified, compared with only 1–21% by volume in Griffiths & Barker's (1966) study, possibly because of a greater proportion of soft plant species available after periods of above-average rainfall.

The major findings

Despite the differences among studies in the techniques used, it is clear that for the large kangaroos *Macropus rufus*, *M. giganteus* and *M. robustus erubescens* grass is an important part of their diet. For instance, Ellis *et al.* (1977) found that at one sampling time when Red Kangaroos had 87% identifiable grass particles, sheep in the same land system had only 7% grass. Red Kangaroos, Euros and sheep all selected grasses and forbs when they were readily available. When pastures deteriorated, sheep selected mainly flat-leaved chenopods whereas the two macropodine species selected mainly grass with varying amounts of flat- and round-leaved chenopods. Euros were even more selective than Red Kangaroos, eating grass even when present at very low levels. Thus potential overlap in diet between macropodids and sheep was greatest under good pasture conditions and least under the poorest conditions. Application of Ellis *et al.*'s (1977) conclusions to all situations is limited by the fact that the study did not include a period of drought. Nevertheless, their conclusion is supported by the studies of Bailey *et al.* (1971) and Griffiths *et al.* (1974), both conducted with Red Kangaroos and sheep during drought conditions.

Even in studies in which more overlap in the preferences of kangaroos and sheep has been found, such as that of Storr (1968), there were however considerable areas of non-overlap, in that Red Kangaroos ate more *Aristida* spp. (spear grass) and much less dicotyledon material than did sheep. Again, while Griffiths & Barker (1966) found that sheep and Red Kangaroos ate the same proportions of grasses, the species composition differed. In dry periods both the sheep and kangaroos ate more non-grass material, but the dicotyledons eaten by Red Kangaroos, Eastern Grey Kangaroos and sheep were not the same. Griffiths & Barker (1966) concluded that the food preferences of the three species were specific enough for the plants of the habitat and the three herbivores to be regarded as an ecosystem.

The conclusion from the results of these detailed studies is that *M. giganteus*, *M. rufus* and *M. robustus erubescens* are all primarily grazers. Other species of the genus *Macropus* also appear to be predominantly grazers, but the evidence is far less complete. *M. rufogriseus* (the Red-necked Wallaby) grazes in open cleared country (Calaby, 1966). The staple diet of *M. parryi* (The Whiptail Wallaby) consists of kangaroo grass (*Themeda australis*), but tips of long bunch spear grass (*Heteroropogon contortus*) are also eaten (Bell, 1973). *M. agilis* (the Agile Wallaby) appears to be an opportunistic feeder, consuming short green grass and long dry spear grass, as well as browsing on *Melaleuca* and other low shrubs (Bell, 1973). *M. dorsalis* (the Black-striped Wallaby) grazes outside but usually close to its scrub refuge areas (Calaby, 1966). *Themeda australis* appears to be a preferred food of *M. parma* (the Parma Wallaby) (Maynes, 1974).

Information on the dietary preferences of the other three extant genera in Sanson's grazing grade, *Peradorcus*, *Lagorchestes* and *Onychogalea*, is scanty. In fact nothing is known of the diet of any *Onychogalea* species (the Nailtail Wallabies). Newsome (1975) assumed that *Lagorchestes conspicillatus* (the Spectacled Hare-wallaby) was primarily a grazer from its original habitat of Mitchell grass (*Astrebla pectinata*) plains. The diet of *L. conspicillatus* in its present range of three arid islands off the Western Australian coast was described by Main & Yadav (1971) as a mixture of both mono- and dicotyledonous species. Ride & Tyndale-Biscoe (1962) collected two specimens of *L. hirsutus* (the Rufous Hare-wallaby) during their visit to Shark Bay in Western Australia. One, from Dorre Island, had only grass fragments in rectal faecal samples, whereas there was no grass at all in the specimen from Bernier Island, but instead unidentified dicotyledonous material and malvaceous hairs despite the abundance of grasses such as spinifex on this island. A recently discovered colony of *L. hirsutus* in the Tanami Desert in the Northern Territory appeared to be feeding mainly on grasses (both leaves and seed heads), but also on chenopods where these were abundant on salt flats (Bolton & Latz, 1978). From this limited evidence *Lagorchestes* appears to be a grazer or mixed feeder.

Peradorcus concinna (the Little Rock-wallaby) is unique in both its dentition and diet. Although Sanson (1976, 1978) classified it as a grazer, it feeds primarily on the fern *Marsilea* which has a low fibre content but is very high in silica, approximately 26% on a dry-matter basis (G. D. Sanson, personal communication). Its molar morphology puts it into the browsing grade, a system adapted to crushing low fibre plants, but the problem of wear from the high silica content has been resolved by adding teeth. This is done by molar progression, the sequential anterior movement of molars through the dental mill, after which they are lost (Sanson & Miller, 1979). Molar progression occurs in all species of the genus *Macropus*, and so is more characteristic of grazers than browsers. In *Macropus giganteus* it appears to be a device for improving the efficiency

166   6 *Diet and nutrition of kangaroos and wallabies*

of tooth use by maintaining a sharp cutting edge on two molars in the area of occlusion, rather than for reducing tooth wear; *M. giganteus* uses a total of four molars in its life-time, the same number used by the browser *Wallabia bicolor* in which molar progression is absent. In *Peradorcus* molar progression produces extra molars, and in this way is an antiwear device. The position of *Peradorcus* in Sanson's (1978) classification is therefore likely to change.

Among the members of Sanson's (1978) browsing grade we know something of the diets of *Wallabia* and *Setonix*. The study of Harrington (1976) clearly demonstrated that, at least in her study area in north-eastern New South Wales, the Swamp Wallaby (*Wallabia bicolor*) selects the forb and shrub component and not the grass component of the vegetation on offer. She also found that Swamp Wallabies eat small amounts of a wide variety of species, something which is characteristic of eutherian browsers (Jarman, 1973). Edwards & Ealey (1975) reported very similiar findings for *W. bicolor* in Victoria. *Setonix brachyurus*, the Quokka, has been the subject of intense study on its island habitat of Rottnest Island, close to Perth in Western Australia. Little has been done on the remnant mainland population in the south-west corner of the state. Storr (1964a) examined

Plate 6.2. *Dendrolagus matschiei*, a tree-kangaroo from New Guinea. (Ray Williams.)

the diet of Quokkas at three locations on Rottnest, and found that succulents dominated the diet in areas where surface water was never available (79% in spring to 97% in the summer drought). This value was still 52–77% in an area where seepage water was available. The next major dietary component in summer was shrubs, up to 28% in one area. Forbs and grasses only assumed importance in winter and early spring; these were all annual species responding to the winter incidence of rainfall (annual average 74 cm).

Dendrolagus spp. (the tree-kangaroos) (Plate 6.2) are assumed from their habitat of tropical rainforest to feed predominantly on tree foliage. Troughton (1965) reported that along with leaves of a wide variety of trees *D. bennettianus* (Bennett's Tree-kangaroo) ate creepers, ferns and fruits. *Lagostrophus fasciatus* (the Banded Hare-wallaby) is reported to feed on succulents such as *Carpobrotus*, together with some grasses (Main & Yadav, 1971). Ride & Tyndale-Biscoe (1962) collected rectal samples from seven *L. fasciatus* from Dorre Island and ten from Bernier Island. Faeces from all seven Dorre animals contained grass, the mean being 66% grass epidermis and the remainder unidentified plant tissue. In contrast, only three animals from Bernier Island had any grass in their faeces. Of the remainder, two had *Carpobrotus* epidermis and all had malvaceous hairs. The low incidence of grass in the Bernier animals was considered by Ride & Tyndale-Biscoe (1962) to be surprising in view of the abundance of grasses, particularly spinifex, on the sand dunes of that island.

Although the digestive physiology and metabolism of *Thylogale thetis* has been studied in some detail, little is known about the natural diet of this and other *Thylogale* species. Troughton (1965) reported that pademelons both browse and graze, but the extensive changes in the vegetation cover throughout the range of *Thylogale* since the arrival of European man make definite conclusions impossible. Johnson (1977) observed *T. thetis* browsing shrubs such as *Helichrysum diosmifolia* throughout the year, and *Solanum mauritianum* (wild tobacco) during winter months. Grasses were also eaten in the rainforest, and comprised the bulk of the food items taken an open pasture areas at night. Little can be said about the diet of the forest-dwelling *Dorcopsis* and *Dorcopsulus*, but since they are restricted to undisturbed rainforest (George, 1979) the opportunity for feeding on grass is limited. Sanson (1980) quoted two personal communications suggesting that *Dorcopsis* is a browser, taking softer vegetation, flowers and fruits.

Of the species of *Petrogale*, Sanson's (1978) intermediate grade, only the Yellow-footed Rock-wallaby (*P. xanthopus*) has been investigated. Dawson & Ellis (1979) found from faecal analysis that under good conditions the largest component (42–52%) of the diet of *P. xanthopus* was forbs, mostly small ephemeral species. During drought conditions browse became the most important dietary component (44% of intake). Grass constituted between 22 and 34% of the diet. Thus *P. xanthopus* appears to be an example of a mixed feeder.

In summary then, from the information we have on dietary preferences of the extant macropodines, with the exception of *Peradorcus* there appears to be good support for Sanson's (1978) classification of the Macropodinae into the browsing, intermediate and grazing grades. On present evidence *Peradorcus* should perhaps be placed in the intermediate grade (Table 6.1) along with *Petrogale* rather than with the grazers. Of interest here is the present uncertainty about the validity of assigning *Peradorcus* to a genus separate from *Petrogale* (Poole, 1979). Our knowledge of the ecology and particularly the nutritional requirements and dietary preferences of *Petrogale*, *Peradorcus* and many rare species is far from complete. This unsatisfactory state is likely to remain so into the foreseeable future because of the difficult nature of the work. It is seen as a major limitation to the interpretation of findings not only on dental morphology but also on other aspects of digestive physiology as well.

Nutrition and ecology

A number of the more complete studies will now be considered in order to explore in greater detail the role of some of the digestive and metabolic adaptations of the Macropodinae in their ability to exploit different nutritional environments. The constraints of energy, protein and water in each environment to be considered will be central to the discussion.

The arid-zone kangaroos – the Euro and the Red Kangaroo
The Euro

It is appropriate to begin this section with the Euro (see Plate 6.3) because of the earlier comprehensive field studies of Ealey (1967a, b) and Ealey & Main (1967) on this species. More recent work on the nutritional requirements and digestion and metabolism in the Euro has confirmed many of Ealey's predictions, and we now have a good understanding of the ways in which the Euro survives in and exploits its often very harsh habitat.

Ealey's (1962) study area was the Pilbara district in the north-west of Western Australia. It is an area of leached-out soils and poor vegetation. Low rainfall, low humidity and very high summer temperatures produce

Plate 6.3. *Macropus robustus erubescens*, the Euro, from western New South Wales. (Ray Williams.)

a harsh environment in which, in aboriginal times, the numbers of Euros were low and their distribution patchy. The long hot summers and lack of surface water would have induced severe annual mortality, and restricted hardy survivors to rocky hills. Here they would have found caves in which to escape the desiccating heat and patches of *Triodia pungens* (spinifex) on which to feed.

Ecology. The introduction of domestic sheep by European man into the Pilbara in 1866 (Ealey, 1967a) disrupted the pattern of annual mortality in the Euro population. The sinking of dams and wells every 5–8 km meant that water no longer acted as a population control on the Euro (Ealey, 1967b). Overgrazing by sheep meant that the more nutritious native grasses were soon eaten out and the durable spinifex encroached onto the plains, eventually to dominate them. Spinifex is a submaintenance diet for sheep, but not for the Euro with its low maintenance protein requirement (Brown & Main, 1967). Consequently Euro numbers increased along with the sheep population. By the late 1920s sheep numbers had increased to nearly 800000 but a drought in 1935–36 caused a drastic reduction in numbers, and numbers have since continued to fall. Meanwhile the Euro continued to thrive, becoming more abundant than the sheep. Today, Euros even inhabit the plains, sheltering under trees by day like Red Kangaroos. The additional water lost in cooling is readily recouped at stock-waters that were so important in disrupting the old ecological regime in the first place (Newsome, 1975).

Ealey & Main (1967) studied the seasonal cycle in the nutritional status of two Euro populations in the Pilbara, one where there was only low protein forage available (Mt Edgar), the other where a mixed vegetation existed containing plants of high and low protein content (Woodstock). Although Euros are able to live and even to breed on low protein vegetation, many thin and starving animals were seen by Ealey & Main (1967) during periods of protein shortage even though there was an excess of edible vegetation. After a prolonged dry season spectacular mortality may occur. The animals which died were always emaciated. Thus it seems probable that nutrition now plays a major role in regulation of Euro density in areas of adequate water supply in the Pilbara.

Two types of mortality were observed by Ealey & Main (1967). The first was the regular seasonal die-off of a certain proportion of the population at Mt Edgar where only the hardiest animals were able to survive semi-starvation almost every summer. Mt Edgar Euros consistently had lower blood haemoglobin levels in summer than did Woodstock Euros, reflecting the low protein status of the vegetation (Fig. 6.2).

Nutrition and ecology

The second type of mortality was the population crash seen in the Woodstock Euros after a particularly dry summer. In most years early summer storms and the subsequent monsoonal rains resulted in Woodstock Euros being in a good nutritional state to survive the latter part of the summer and autumn. A number of such years resulted in a gradual increase in the population, with little selection pressure against less well-adapted individuals. When the Euro density became high the more nutritious vegetation would be eaten very early in the season. The animals of this dense population could not attain a sufficiently good state of nutrition to last them over a dry summer. A prolonged dry season would result in a spectacular population crash, as occurred in 1954 (Ealey & Main, 1967).

Physiology. Notwithstanding such population crashes, compared with the sheep the Euro must be regarded as being extremely well adapted to an arid environment. The physiological bases of this adaptation have been

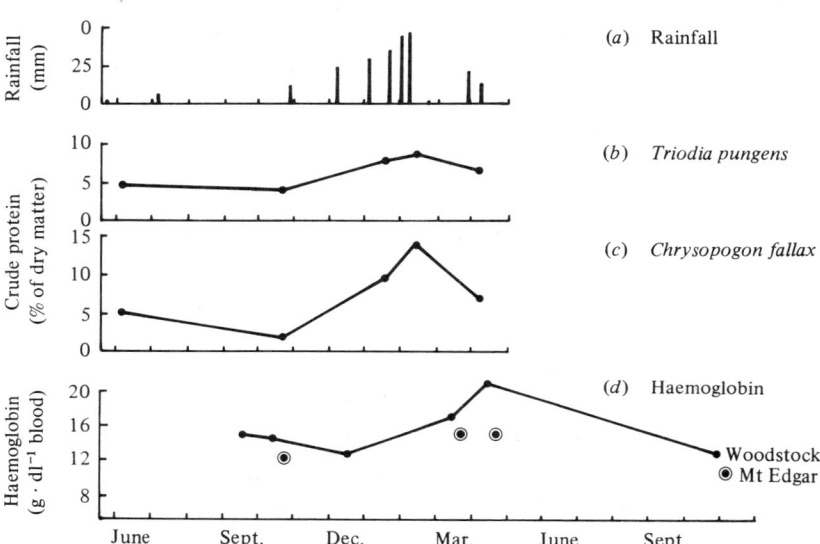

Fig. 6.2. Changes in the protein content of (*b*) *Triodia pungens* (spinifex) and (*c*) *Chrysopogon fallax* in the Pilbara region in response to rain (*a*). Note the difference in protein content of the two species in summer. Mt Edgar vegetation consists mainly of *Triodia*; this is reflected in the lower haemoglobin concentrations in blood (*d*) of Mt Edgar Euros compared with Woodstock Euros. After Ealey & Main (1967).

reasonably well researched. The first of these is related directly to the generally low basal metabolic rate of marsupials (Chapter 1). Thus the adult Euro has a maintenance energy requirement 27% below that of sheep (Hume, 1974). Its maintenance nitrogen requirement of 160 mg truly digestible nitrogen per $kg^{0.75}$ per day (Brown & Main, 1967) is the lowest of the marsupials so far examined, and less than half that of eutherians such as the sheep and horse (Table 1.7). Its rate of excretion of creatinine is also low compared with the few other macropodids that have been examined (Table 1.8), and again less than half that of eutherians. Similarly, the Euro has a water turnover rate under non-stress conditions approximately 22% below that of the eutherian mean (Denny & Dawson, 1973 and Table 1.9), although not different from many other marsupials studied.

These comparatively low nutrient requirements are all factors contributing to the ability of all arid-zone macropodine species to survive long periods of poor nutrition. The efficiency with which the Euro utilises and retains both nitrogen and sulphur when fed poor quality roughage was demonstrated by Hume (1974) in a comparison with Red Kangaroos and sheep (Table 6.2). All three species were in positive balance with respect to both elements on the good quality chopped lucerne hay diet. On the medium quality chopped oaten hay, however, the Red Kangaroos were unable to conserve enough nitrogen or sulphur and slipped into negative balance. On the poor quality milled wheaten straw diet all three species were in negative nitrogen balance, but both the Euro and the sheep remained in slight positive sulphur balance. The wheat straw contained less than 2% crude protein, even lower than the spinifex analysed by Ealey & Main (1967) during the dry season in the Pilbara. Fibre digestibility was higher in the Euro than in the Red Kangaroo on both the lucerne and oaten hay diets, although there was no difference between the two macropodine species on the straw diet.

These results serve to illustrate how well the Euro is physiologically adapted to utilise low protein, high fibre forages such as spinifex. In addition to Hume's (1974) study a number of other investigations have compared the Euro with the Red Kangaroo, the other large arid zone macropodine.

The Red Kangaroo

In contrast to the sedentary Euro, *Macropus rufus* is a highly mobile species of the open plains. Frith (1964), Newsome (1965) and Bailey (1971) have all recorded the movements of Red Kangaroos to areas of better quality feed, and in response to storms 10–20 km distant. Thus it appears that whereas the Euro is physiologically adapted to survival in arid regions where the quality of available forage is often low, the Red Kangaroo owes its survival in the arid zone more to a behavioural adaptation.

Ecology. Newsome (1975) studied a Red Kangaroo population near Alice Springs in the Northern Territory over a period which covered the longest drought on record in central Australia. This drought began in 1958 and lasted until 1966. Red Kangaroos were relatively uncommon in the Alice Springs area before Europeans settled there. The two limiting factors were probably water and protein, just as in the case of the Euro in the Pilbara. Their drought refuges were the flood-outs at the ends of creeks. Here the Red Kangaroos found almost continuous shade under dense stands of trees, and persistent green grass. After widespread good rains, when green grass would have been widely abundant, they could disperse and breed. During drought however, few would have survived outside these drought refuges. European man released ruminant stock onto the grassy plains in

Table 6.2. *The intake and retention of nitrogen and sulphur of three diets by Euros, Red Kangaroos and sheep*

	Chopped lucerne hay	Chopped oaten hay	Milled wheaten straw
Nitrogen			
Intake ($g \cdot kg^{-0.75} \cdot d^{-1}$)			
Euro	1.51	0.50	0.11
Red Kangaroo	1.46	0.43	0.06
Sheep	2.56	0.64	0.14
Balance ($g \cdot kg^{-0.75} \cdot d^{-1}$)			
Euro	+0.37	+0.14	−0.19
Red Kangaroo	+0.14	−0.12	−0.57
Sheep	+0.66	+0.11	−0.24
Sulphur			
Intake ($mg \cdot kg^{-0.75} \cdot d^{-1}$)			
Euro	102	50	34
Red Kangaroo	104	43	20
Sheep	173	53	43
Balance ($mg \cdot kg^{-0.75} \cdot d^{-1}$)			
Euro	+11	+9	+1
Red Kangaroo	+1	−3	−38
Sheep	+12	+14	+2

After Hume (1974).

the 1870s, and sank wells and dams every 5–15 km across them to provide his stock with water. This removed one factor limiting the size of the Red Kangaroo population. The cropping of the standing long dry grass by domestic stock created a subclimax grassland of generally shorter but greener, higher protein feed. This largely removed the second factor limiting the Red Kangaroo population. This meant that breeding could be prolonged with greater success, and more kangaroos could survive periodical droughts. The result was an enormous increase in Red Kangaroo numbers throughout the arid zone. However, more recently there has been a continuous decline in Red Kangaroo numbers in north-western Australia, including the Pilbara district (Ealey, 1967a), presumably as a result of overgrazing of the more nutritious pasture species by Red Kangaroos, Euros and domestic stock. The accompanying encroachment of spinifex onto the open plains in the Pilbara has excluded both sheep and the Red Kangaroo from much of that area. Only the Euro, with its physiological capacity to utilise the coarse spinifex, has continued to thrive (Newsome, 1975).

Reproduction. The significance of the abundance and quality of food to the Red Kangaroo is emphasised by the relationship in Fig. 6.3 between

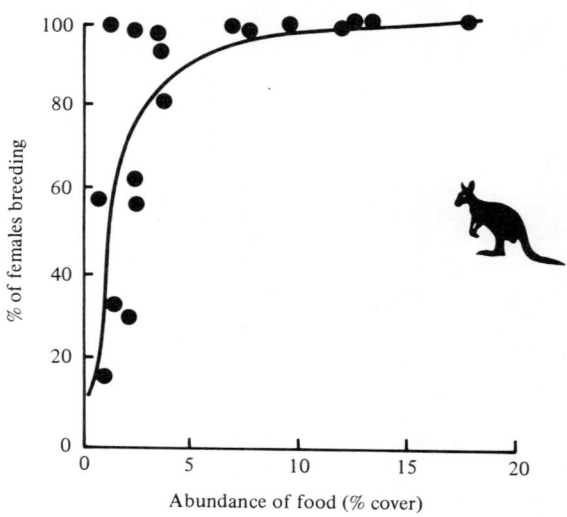

Fig. 6.3. Relationship between the percentage of adult female Red Kangaroos in breeding condition and the abundance of food. After Newsome (1966).

the percentage of adult females in breeding condition and the abundance of food (Newsome, 1966, 1971).

In addition to a failure of post-partum ovulation during drought (Newsome, 1964), other effects of reduced food supply on the Red Kangaroo have been documented by Frith & Sharman (1964). There was some intrauterine mortality but the principal mortality occurred among pouch young at the end of pouch life, from 196 days of age onwards, and among the young-at-foot. The mortality was greatest in drought areas where 83% failed to reach maturity.

Physiology. Returning now to the physiological comparisons between the Red Kangaroo and the Euro, the data of Denny & Dawson (1975a) suggest no significant difference in water turnover rate in non-stress situations between the two species. Similarly, both species were 'camel-like' in their response to dehydration (when there was a 20% reduction in body weight) in that plasma volume was maintained, falling by only 8.3% in Red Kangaroos and 7.4% in Euros (Denny & Dawson, 1975b). Animals that are capable of maintaining a constant plasma volume have a greater chance of survival during dehydration because of the effects on blood viscosity (Schmidt-Nielsen, 1964). Of the large mammals that have been studied only two can maintain plasma volume during dehydration as effectively as the Euro and the Red Kangaroo. These are the camel, which loses less than 10% of its plasma volume during a 20% weight loss (Schmidt-Nielsen, 1964) and the burro (*Equis asinus*), which loses 7% of its plasma volume during a 20% weight loss (Yousef, Dill & Mayes, 1970). Other mammals which have been studied lose from 40 to 50% of their plasma volume for a similar level of dehydration (Macfarlane *et al.*, 1961; Rosenmann & Morrison, 1963).

The pattern of water loss from other body compartments differed between the two macropodine species, particularly gut water loss. This compartment contributed 56% of the total water loss of Red Kangaroos but only 22% of the loss from Euros. The preferential maintenance of gut water in the Euro at the expense of interstitial water could give that species a significant advantage in maintaining the fermentation in the forestomach during drought. This aspect of gut function has not been investigated.

Also of significance is the finding by Denny & Dawson (1977) that Euro kidneys reabsorbed much more urea from the glomerular filtrate when dehydrated (89%) than did Red Kangaroo kidneys (69%). As a consequence plasma urea concentration in Euros was almost twice that of Red

Kangaroos when both species were dehydrated to 14% reduction in body weight; in hydrated animals plasma urea concentrations were similar between the species. Although not measured, recycling of urea to the gut in the dehydrated Euro would be expected to be significantly greater than in the dehydrated Red Kangaroo.

The combination of the preferential maintenance of gut water and the expected greater urea recyling in the Euro under conditions of water shortage must be of great ecological significance. Since Euros are much less mobile than Red Kangaroos the continued efficient functioning of the digestive system during drought when only poor-quality feed is available must be important to the long-term survival of this sedentary species.

The ability of the Euro to reabsorb more filtered urea than the Red Kangaroo was correlated with the relative medullary thickness (RMT) of their kidneys which was 7.2 in the Euro and 5.8 in the Red Kangaroo (Dawson & Denny, 1969). The significance of the RMT as an index of the concentrating ability of the kidney and as a measure of the animal's ability to withstand dehydration was discussed in Chapter 3. The RMTs of a range of macropodid marsupials are shown in Table 6.3. Although there is a marked difference in RMT between *Dendrolagus matscheii* from New Guinea at one end of the spectrum and *Lagorchestes conspicillatus* and *Bettongia lesueur* from Barrow Island off the north-west Australian coast at the other end (Yadav, 1979), within the range there is considerable overlap between habitat types. Thus minor differences in RMT must be interpreted with some caution. For instance, despite its lower RMT the Red Kangaroo has better urine-concentrating abilities than the Euro (Dawson & Denny, 1969). This may be related to the greater heat loads and thus a greater need to conserve water in the open plains-dwelling Red Kangaroo than in the Euro which traditionally shelters from the summer heat in caves and under rock ledges. Dawson & Brown (1970) also found that the Red Kangaroo has fur which gives greater protection from solar radiation in summer, and from heat loss in winter, than does that of the Euro. This difference can again be related to the contrasting micro habitats of the two species.

The island wallabies – the Quokka and the Tammar Wallaby
The Quokka

Setonix brachyurus was formerly abundant throughout the south-west of Australia. Following a catastrophic decline in the 1930s the mainland population was restricted to densely vegetated swamps where the Quokka still occurs in low numbers. There are two extant island populations, one on Bald Island off the southern coast and the other on Rottnest Island near Perth. The Quokka remains common on the former island and abundant on Rottnest.

Ecology. The Quokka on Rottnest Island, freed from competition and predation after the island was isolated from the mainland some 7000 years ago, has undergone an expansion of its niche. The diet of the mainland Quokka of sedges, perennial herbs and leguminous and myrtaceous shrubs (Storr, 1964b) is not subject to fluctuations in fibre, nitrogen and water content to the same extent as that of the Rottnest Island Quokka. Initially most of the habitats occupied by the Rottnest Quokka were heavily vegetated with *Callitris*, *Melaleuca* and *Acacia* shrubs (Storr, 1964a). The advent of human settlement on the island led to marked changes. Much of the shrub vegetation was cleared or thinned, which enabled herbaceous plants, particularly exotic annuals, to flourish. Frequent burning and severe browsing of regenerating shrubs led to the present situation in which much of the island is now dominated by the unpalatable shrub *Acanthocarpus* and the perennial grasses *Poa* and *Stipa* spp. In winter, germination of annuals and new growth of perennials provides a high-quality diet, but in summer only perennial shrubs and grasses provide food. The perennial shrubs become severely over-browsed and the Quokkas are forced to eat more of the fibrous perennial grasses and the older, more fibrous leaves and twigs of perennial shrubs.

This suggests that there may be differences between the mainland and island Quokkas populations in aspects of their biology related to nutrition. Indeed, Shield (1956) showed that the mainland population breeds continuously throughout the year; in contrast, Rottnest Quokkas have a season of birth restricted to six months. When female Quokkas from Rottnest were maintained in captivity on good quality feed *ad libitum* however, the annual anoestrus period decreased, and after a period of two years or more finally disappeared. Since the climate of the two areas, only 80 km apart, is similar, the breeding season difference between the two populations thus must be related to food supply.

Studies on adrenocortical function suggested that the Rottnest Quokka underwent a seasonal stress (Herrick, 1961; Miller & Bradshaw, 1979), although the latter workers were able to conclude that the seasonal mortality of the Quokkas on Rottnest Island did not result from any breakdown in adrenal function, despite a substantial decline in condition of the animals. Shortages of digestible energy, nitrogen and water have variously been implicated, together with salt loading (Shield, 1959; Storr,

1964a; Barker, 1974). It may well depend upon which part of the 1900 hectare island is under consideration. For instance, one population of Quokkas lives on West End (Fig. 6.4) which entirely lacks fresh water. Others, living in the Lakes Area, have access to fresh water seepages around the salt lakes throughout the year. No Quokka marked on West End has ever been recorded on the other side of a 200-metre-wide isthmus between the two areas. This leads to marked differences between the two populations in their access to free water and in the quality of their diet (Storr, 1964a). West End animals show much greater fluctuations between winter and summer in haemoglobin concentrations and in body condition than animals from the Lakes Area (Tyndale-Biscoe, 1973).

Table 6.3. *Relative medullary thickness (RMT) of macropodid marsupials*

Species (and number of animals)	RMT	Habitat	Reference
Dendrolagus matschiei (1)	3.9	Tropical rainforest	Yadav (1979)
Thylogale thetis (4)	5.7 ± 0.2	Temperate rainforest	Hume & Dunning (1979)
Setonix brachyurus	5.1–5.8	Mediterranean, summer drought	Brown (1964)
Macropus irma (1)	5.8	Mediterranean, summer drought	Yadav (1979)
Bettongia penicillata (2)	5.8, 5.9	Mediterranean, summer drought	Yadav (1979)
Macropus eugenii (19)	5.9 ± 0.3	Mediterranean, summer drought	Kinnear et al. (1968)
Macropus giganteus (1)	5.3	Mediterranean, summer drought	Yadav (1979)
Macropus rufus	5.8	Semi-arid to arid	Dawson & Denny (1969)
M. robustus erubescens	7.2	Semi-arid to arid	Dawson & Denny (1969)
Peradorcus concinna (1)	7.3	Semi-arid, winter drought	Yadav (1979)
Macropus eugenii (18)	6.8 ± 0.3	Maritime arid	Kinnear et al. (1968)
Lagorchestes hirsutus (2)	6.0, 7.2	Maritime arid	Yadav (1979)
Lagostrophus fasciatus (1)	7.5	Maritime arid	Yadav (1979)
Bettongia lesueur (2)	8.0, 8.8	Maritime arid	Yadav (1979)
Lagorchestes conspicillatus (2)	8.4, 9.0	Maritime arid	Yadav (1979)

Nutrition. In an attempt to define the limiting factor(s) to the West End Quokkas during summer and autumn Wake (1980) conducted both short- and long-term supplementation experiments. Two short-term experiments were based on the technique used by Kinnear (1970) in his work with *Macropus eugenii* on the Abrolhos Islands. The treatments were control, urea supplemented, starch supplemented, and both urea and starch supplemented. The urea (200 mg of nitrogen in the early summer experiment, 93 mg in the late summer experiment) was dissolved in 2 ml distilled water and injected intraperitoneally. The starch (20 g suspended in 20 ml water in both experiments) was administered by stomach tube. Animals were caught just after dark, weighed, and an initial blood sample taken; the early summer experiment involved 21 animals, and the late summer experiment 24. All animals were injected intraperitoneally with tritiated water to determine the size of the body water pool.

After receiving one of the four treatments the animals were kept for 12 hours in individual metabolism cages and urine and faeces collected separately. At the end of 12 hours a second blood sample was taken and the animals were released at the sites of capture. Changes in plasma urea concentrations and pool sizes over the 12-hour period were used to calculate a value for apparent utilisation of urea on each treatment. The results from the two experiments are summarised in Table 6.4.

The conclusions from the experiments were that in early summer energy primarily and nitrogen secondarily were limiting microbial activity in the Quokka forestomach; the greatest utilisation of injected urea occurred

Fig. 6.4. Rottnest Island, showing the locations of West End and the Lakes Area, the sites of several field studies of *Setonix brachyurus*.

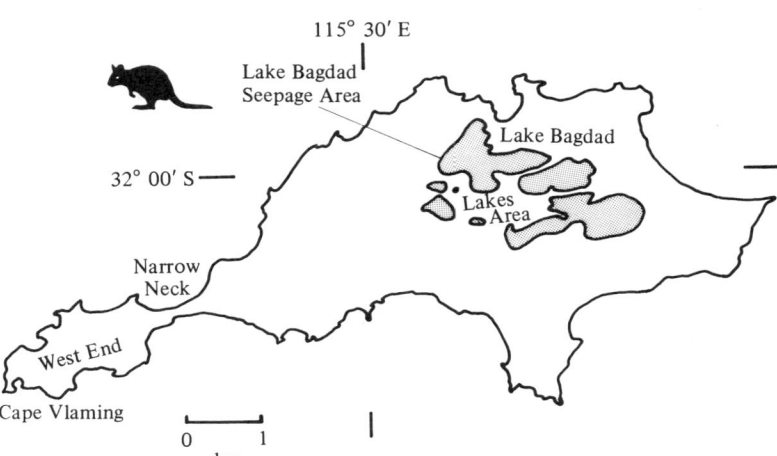

when both energy and nitrogen were supplemented. In late summer energy was the limiting factor, as shown by the negative increments in plasma urea nitrogen concentrations when starch alone was given. Presumably the difference between the two experiments was due to the fact that in the early summer the Quokkas were still in good condition and the food plants were not yet severely deficient in either readily fermentable carbohydrate or protein. As the summer progressed the decline in the level of readily fermentable carbohydrate in plants apparently assumed greater significance than the decline in protein level.

Wake (1980) then conducted long-term supplementation experiments in mid to late summer by distributing through part of the study area either starch, casein or water, each in a different year. Each supplement was mixed with a fluorescein dye, which is quantitatively excreted in the urine, in order to identify those animals which took the supplement, and to obtain an estimate of the amount of supplement eaten. Wake (1980) also developed an index of body condition, the Condition Factor, which was the ratio of actual body weight to predicted body weight. Predicted body weight was derived from an allometric equation relating body weight to leg length. Thus a change in Condition Factor meant that the same proportion of the animal's weight had been gained or lost regardless of size. In Wake's (1980) study the extremes of Condition Factor were 0.64 and 1.45.

Table 6.4. *Effect of short-term (12-hour) supplementation with urea, starch or both on urea utilisation in Quokkas at West End, Rottnest Island*

	Treatment			
	Control	Urea	Starch	Urea + Starch
Early summer				
Plasma urea (mg N·dl^{-1}), T_0	16.0	20.8	24.0	19.6
Plasma urea (mg N·dl^{-1}), T_{12}	27.3	32.8	28.9	27.0
Increment	11.2	12.0	4.9	7.4
Apparent utilisation of urea N (mg)	—	86	65	211
Late summer				
Plasma urea (mg N·dl^{-1}), T_0	21.7	20.4	20.4	22.5
Plasma urea (mg N·dl^{-1}), T_{12}	28.3	30.5	18.7	22.1
Increment	6.6	10.2	−1.7	−0.4
Apparent utilisation of urea N (mg)	—	50	141	175

After Wake (1980).

Nutrition and ecology

The energy supplement was taken readily. A total of 92 kg of starch was made available to the population, and almost all was consumed by Quokkas – although other nocturnal animals such as rats and mice may also have consumed some starch. The supplements were always distributed just after dark and any excess was removed before sunrise to prevent losses to birds. A total of 156 Quokkas was caught over six catching trips spanning a 14-week period between commencement of supplementation and soon after the beginning of the winter rains. On each occasion

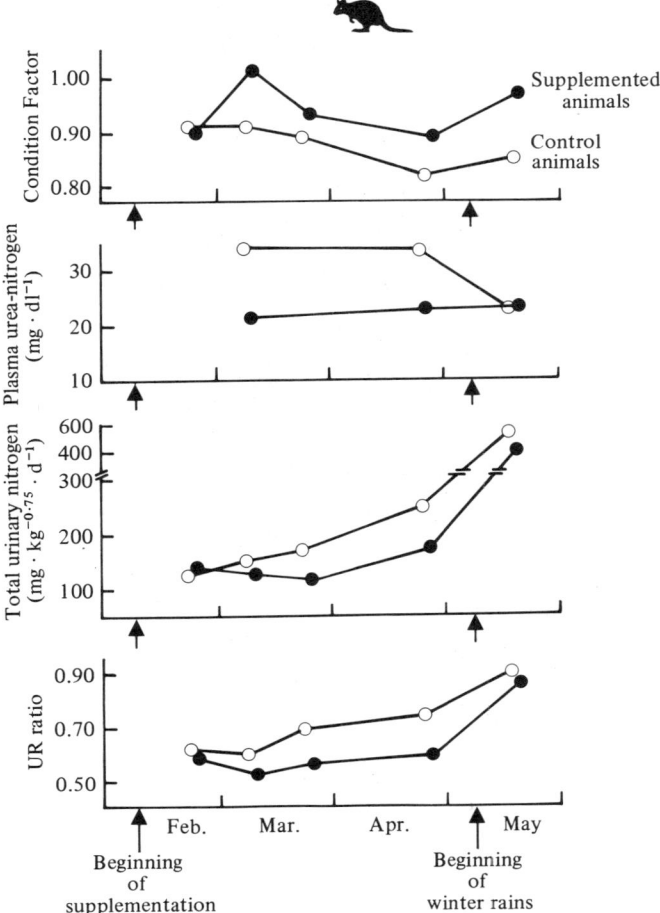

Fig. 6.5. The response of *Setonix brachyurus* to long-term energy supplementation at West End. Condition Factor is the ratio of actual body weight to predicted body weight. Predicted body weight is derived from an allometric equation relating body weight to leg length. After Wake (1980).

supplemented and unsupplemented animals were kept in individual metabolism cages for 12 hours for urine and faeces collections as before. A summary of results appears in Fig. 6.5.

The significantly higher Condition Factor of the Quokkas consistently eating the starch supplement and the increase in Condition Factor for some of those animals shows that readily fermentable energy was an important limiting factor in the diet of the Quokka at West End during mid to late summer. This conclusion is supported by the observed differences in the other three parameters in Fig. 6.5 upon starch supplementation.

The long-term protein supplementation experiment was not successful. Only 6 of a total of 94 animals caught over four catching trips between late December and late March had high urinary fluorescein concentrations, confirming the observation that few animals were eating significant amounts of the supplement despite the use of attractants such as vanilla essence and aniseed oil and of agar and water to increase its palatability. The water supplement was taken more readily. Quokkas were frequently observed drinking from the water containers throughout the ten-week supplementation period, even during rain. The rain which fell just after water supplementation began would have alleviated any dehydration of both supplemented and unsupplemented populations and reduced the effects of the water supplement. However, results from 33 animals caught over two catching trips after supplementation commenced suggested that dehydration was a contributing factor, though not the primary cause, of weight loss in Quokkas at West End in late summer. Condition Factor increased with water supplementation, but not as much as after rain. Parallel increases in total body water (Fig. 6.6) indicate that the improvement in condition represented rehydration rather than an increase in body solids.

Storr (1964a) and Barker (1974) considered that the symptoms of starvation exhibited by Quokkas in summer were a secondary effect of water deprivation. Water restriction reduced voluntary food intake in the Tammar (Barker, Lintern & Murphy, 1970). Storr (1964a) argued that the absence of free water at West End forced Quokkas to eat such a high proportion of succulents to satisfy their water requirements that the high water content of the succulents reduced their dry matter consumption. However, this need not necessarily lead to a reduced digestible energy intake since the carbohydrates of succulents are generally more digestible than those of perennial shrubs and grasses. Also, Quokkas from the Lakes Area, with access to free water, still undergo seasonal anaemia and weight loss (Shield, 1959; Barker, Glover, Jacobsen & Kakulas, 1974). Thus other factors such as a deficit of energy, and perhaps protein also, must be more important.

A further factor contributing to the seasonal debility of Rottnest Quokkas, not directly examined by Wake (1980), may be salt loading. Ramsay (1966) found that in field-caught Quokkas held in metabolism cages, 12-hour urine volumes were always higher for 'coastal' (West End) animals than for 'inland' animals. Except in spring, the latter always showed lower urinary sodium levels despite the lower urine output, suggesting that the West End Quokkas were salt loaded in late summer, which would tend to exacerbate any dehydration resulting from a lack of free water. Ramsay (1966) interpreted her results as being due in part to

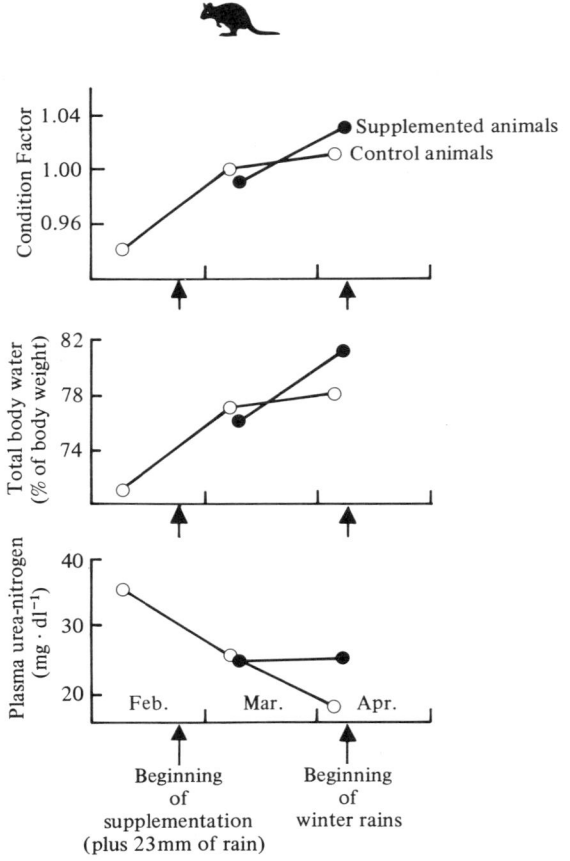

Fig. 6.6. The response of *Setonix brachyurus* to long-term water supplementation at West End. After Wake (1980).

the incidence of salt spray on plants in coastal situations, and in part to the high salt content of succulents eaten by Quokkas at West End.

Thus a number of possible causes have been proposed for the seasonal stress experienced by the Quokka on Rottnest Island. As suggested by Ramsay (1966), the primary cause may well differ between the Lakes Area and West End, despite the similar nature of the seasonal anaemia which always develops (Barker *et al.*, 1974). Thus the conclusion from the short-term supplementation experiments of Wake (1980) that in early summer Quokkas on West End were affected primarily by a shortage of digestible energy need not necessarily apply to the Lakes Area. Similarly, on West End the primary limiting factor may well differ between seasons and between years. It would seem that the most important factors involved are the levels of protein and readily fermentable carbohydrate in the plant material consumed.

The Tammar

Macropus eugenii has been the subject of numerous laboratory investigations, but only comparatively recently has its field ecology received any great attention. Like the Quokka, the Tammar Wallaby was once widespread in south-west Western Australia, and in addition was common in South Australia. It is now largely confined to a number of offshore islands extending from Kangaroo Island (36° south latitude) in the east to the Abrolhos Islands (28–29° south) in the west (Tyndale-Biscoe, 1973). In fact it is on these islands, at opposite ends of its range, where the Tammar has been investigated most extensively. Two field studies will be considered here. The first arose from circumstantial evidence that Tammars on the Abrolhos Islands supplement their water intake by drinking sea water, the second from the need for an effective management plan for the Tammar on Kangaroo Island where it is regarded as a pastoral pest.

The Abrolhos Islands. The first observations of an Australian marsupial were made by the Dutch navigator Francisco Pelsaert, after his ship the *Batavia* was wrecked on Houtman's Abrolhos in 1629; he described the Tammar Wallaby. The climate of the Abrolhos may be considered to be maritime semi-arid (Kinnear, Purohit & Main, 1968). The annual average rainfall is no more than 350 mm, whereas annual evaporation is approximately 1400 mm. For only four winter months does the average rainfall exceed effective rainfall. The long dry summer is characterised by strong and persistent winds. The soils are sandy or rocky and fresh surface water is unavailable for long periods. In this harsh environment the Tammar Wallaby is subjected annually to a prolonged period of poor quality food and a shortage of fresh drinking water.

The first aspect of the Abrolhos Tammar's physiology to be investigated by Kinnear *et al.* (1968) and Purohit (1971) was its ability to survive when the only drinking water available was sea water. Kinnear *et al.* (1968) caught three Tammars at the water's edge on a beach on one side of the Abrolhos Islands. These animals were taken to the University of Western Australia where, seven days after capture and without any acclimation period, they were placed in metabolism cages and offered a maintenance diet of chopped oaten hay supplemented with starch, sucrose, molasses, casein and minerals, and sea water for 30 days. The results were remarkable considering the absence of any acclimation period. A 10–15% weight loss occurred over the first three days, but then weight loss ceased and all animals slowly regained weight. Diarrhoea was never observed. Clearly Tammars can survive for considerable periods by drinking sea water, although, as Purohit (1971) demonstrated, dry food and sea water alone will not support the Tammar Wallaby indefinitely either in the field or in the laboratory.

Additional experiments by Kinnear *et al.* (1968) sought to determine the physiological adaptations involved in the Tammar's ability to drink sea water. When fed either a high- or a low-protein dry diet with 100% sea water the maximum urine condosity (i.e. the molarity of a sodium chloride solution having the same specific conductance) was 1.1, very similar to that in Merriam's Kangaroo-rat (*Dipodomys merriami*), a desert rodent shown by Schmidt-Nielsen & Schmidt-Nielsen (1950) to be able to survive on a dry grain diet with 100% sea water. This indicated that, like *D. merriami*, the Tammar had an excellent capacity for excreting electrolytes economically with respect to water. The absence of diarrhoea on the 100% sea water treatments indicated that Tammars have a high capacity to absorb salt and water from the gut, thus avoiding purgation. Kinnear *et al.* (1968) also suggested that their results indicated that the Tammar has a low minimal water turnover rate. This has been confirmed by Denny & Dawson (1975a); under non-stress conditions these latter workers recorded a water turnover rate of $65 \text{ ml} \cdot \text{kg}^{-0.80} \cdot \text{day}^{-1}$ in the Tammar compared with their mean for five macropodid marsupials of $90 \text{ ml} \cdot \text{kg}^{-0.80} \cdot \text{day}^{-1}$ (Table 1.9).

The urine-concentrating ability of the Abrolhos Tammar was reflected in a high relative medullary thickness (RMT) of its kidneys, 6.8 ± 0.3. This is higher than either the Quokka from Rottnest Island (5.1–5.8)

186   6 *Diet and nutrition of kangaroos and wallabies*

(Brown, 1964) or the Tammar from Garden Island near Rottnest (5.9±0.3, Kinnear *et al.*, 1968) (Table 6.3). This raises the question of physiological differences between geographically isolated populations of the same species. As suggested by the lower RMT, Garden Island Tammars appear unable to concentrate their urine to the same extent as Abrolhos Tammars, since they lost weight on 100% sea water while Abrolhos Tammars maintained weight (Kinnear *et al.*, 1968).

The other aspect of the Abrolhos Tammar examined by Kinnear (1970) was its ability to recycle urea from the blood to the digestive tract when faced with a low protein diet. Body weights of Abrolhos Tammars undergo a cyclic pattern, from a maximum in mid-winter to a minimum in mid-summer (Fig. 6.7). However, by late summer body weights were no different from mid-summer, implying that maintenance of stable body weight by the population had been achieved.

The weight loss in late spring is associated with the onset of the annual drought and a decline in plant protein levels. Concomitantly there was a fall in the urinary UR ratio. The concept of the UR ratio as an index of urea recycling was introduced in Chapter 4 in relation to the low urinary nitrogen loss and high urea recycling rate of Koalas (Cork, 1981). The UR

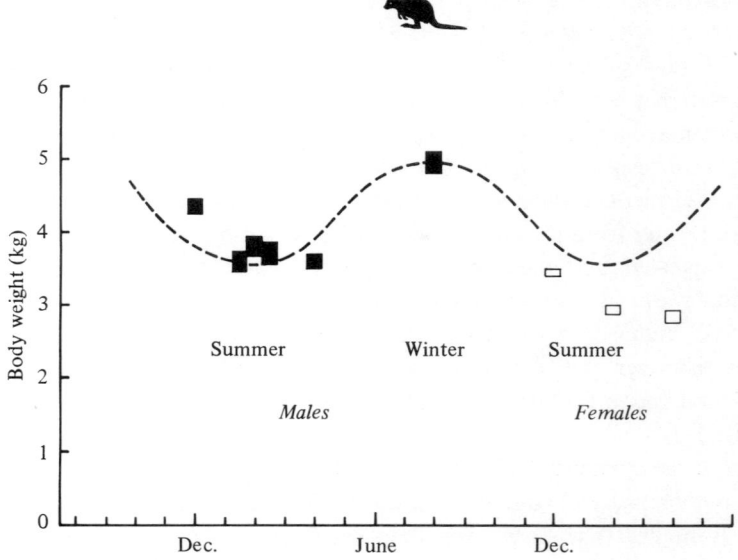

Fig. 6.7. Seasonal changes in body weight of adult *Macropus eugenii* on the Abrolhos Islands in response to the high incidence of rainfall in winter and the annual summer drought. After Kinnear & Main (1975).

Nutrition and ecology 187

ratio is the ratio of urea-nitrogen to total nitrogen concentrations in the urine. In Kinnear's (1970) Tammars the UR ratio was 0.72 in winter, but only 0.30 in summer (Fig. 6.8). This suggested to Kinnear & Main (1975) that in summer, when the animals were protein deficient, urea was being retained by the kidneys, and instead of being excreted some was diffusing from the blood into the gut where a certain proportion was incorporated into microbial protein.

In order to test this hypothesis Kinnear & Main (1975) conducted a laboratory experiment in which one Tammar Wallaby and one Red Kangaroo (*Macropus rufus*) were infused duodenally with an almost nitrogen-free but otherwise complete nutrient mixture. By by-passing the

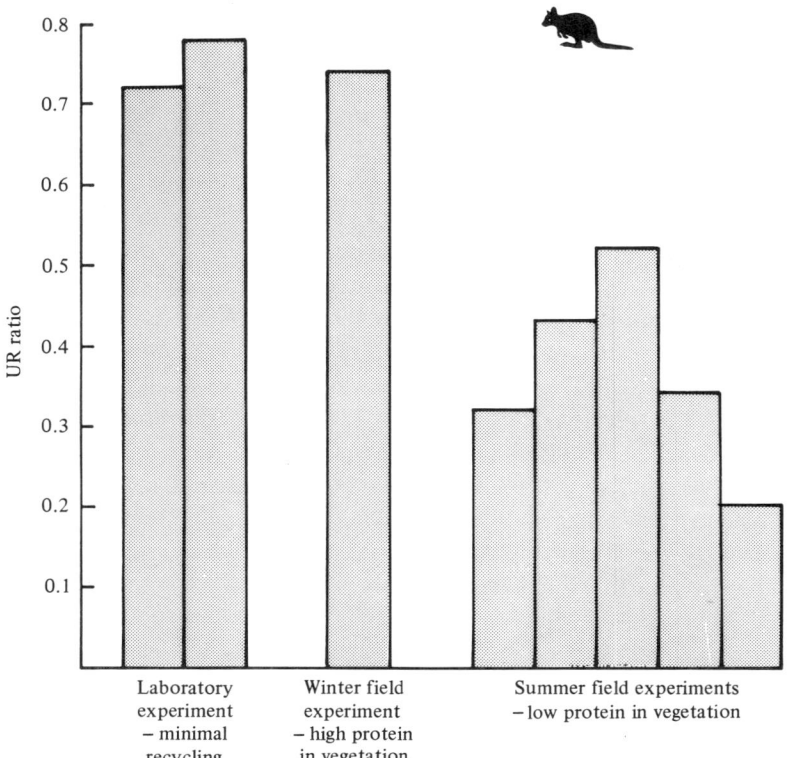

Fig. 6.8. The proportion of total urinary nitrogen in the form of urea (the UR ratio) in *Macropus eugenii* in a laboratory experiment in which recycling of urea from the blood to the digestive tract was minimal, in winter when protein levels in the Abrolhos Islands vegetation is maximal, and in summer when the vegetation is low in protein. After Kinnear & Main (1975).

stomach in this way recycling of urea from the blood was virtually prevented, since the stomach is the major site of microbial protein synthesis in the macropodine digestive tract as we saw in Chapter 5. In these two animals the UR ratio was 0.72 and 0.78, very similar to the mean winter value in the field population (Fig. 6.8). On this basis Kinnear & Main (1975) argued that the Abrolhos Tammars in winter were not recycling significant quantities of urea, presumably because their dietary protein intake was more than adequate to satisfy the nitrogen requirements for maximal microbial growth in the forestomach. However, in summer, from the low UR ratio it appeared that urea recycling made a substantial contribution to the animal's ability to utilise the low protein vegetation available. Support for this contention comes from the stabilisation of body weight of the field population between mid-summer and the beginning of the winter rains (Fig. 6.7), suggesting that the animals must have been close to nitrogen equilibrium during the autumn period.

In another experiment Kinnear & Main (1975) captured Tammars on the Abrolhos Islands at dusk and kept them in individual metabolism cages for 12 hours. Over this 12-hour period the fate of urea injected intraperitoneally was estimated from urinary urea output and changes in the size of the plasma urea pool. The use of this technique by Wake (1980) with Quokkas on Rottnest Island has already been described (page 179). The disappearance from the blood of Tammars of injected urea in early and late summer suggested to Kinnear & Main (1975) that at those times nitrogen was the primary limiting factor in the vegetation. In mid-summer, however, all the injected urea was either detectable in the blood or excreted in the urine. The lack of urea utilisation at this time implies that energy, rather than nitrogen, was primarily limiting. When Kinnear (1970) administered starch by stomach tube to these animals urea utilisation improved dramatically, confirming that the supply of readily fermentable energy to the forestomach microbes was sub-optimal in mid-summer.

Thus deficiencies of both energy and protein have been implicated in the seasonal weight loss of the Abrolhos Tammar population, just as they were in Wake's (1980) study of the Quokka on Rottnest Island. However, because of the Tammar's well-adapted kidneys, dehydration and salt loading appear not to be important.

Kangaroo Island. In contrast to the Abrolhos Islands, Kangaroo Island has an average annual rainfall of 800 mm and a growing season of about 6.5 months. However, the summers are still dry, and the Tammars inhabiting Flinders Chase National Park on the western extremity of the island

Nutrition and ecology 189

exhibit cyclical weight changes. Inns (1980) found that males underwent much greater fluctuations in body weight than females (Fig. 6.9), and he suggested that mating activity by males in late summer contributed to this difference. However, there was considerable loss in body weight of both females and males between early summer and winter; in one year (1978) this amounted to 16% in females and 21% in males. This was reflected in seasonal changes in the amount of fat present within the peritoneal cavity (Fig. 6.10).

Similarly, growth rates of young Tammar Wallabies after leaving the pouch for the last time in spring were rapid, but over the late summer and winter period there was little growth (Fig. 6.11).

For the Kangaroo Island Tammar there is no evidence that nitrogen is a limiting factor at any time of the year as it is for the Abrolhos Tammar. Barker (1971) caught Tammars on Kangaroo Island at various times throughout the year and held them in individual metabolism cages without food or water for 24 hours. Urine and nitrogen excretion were both less in summer than in winter, but the UR ratio was never less than 0.70, indicating that nitrogen deficiency was probably not a problem. Inns (1980)

Fig. 6.9. Seasonal changes in body weight of adult male and female *Macropus eugenii* on Kangaroo Island, South Australia. After Inns (1980).

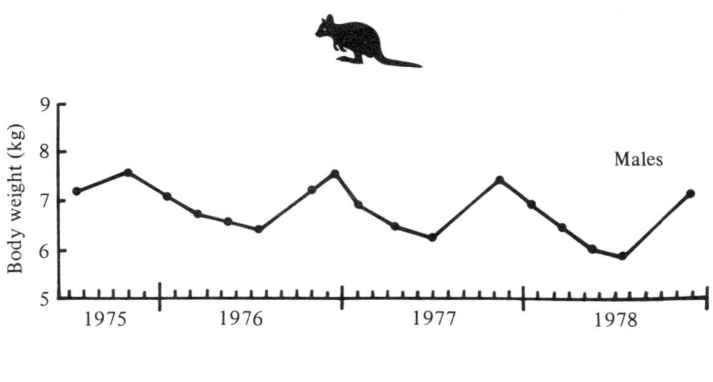

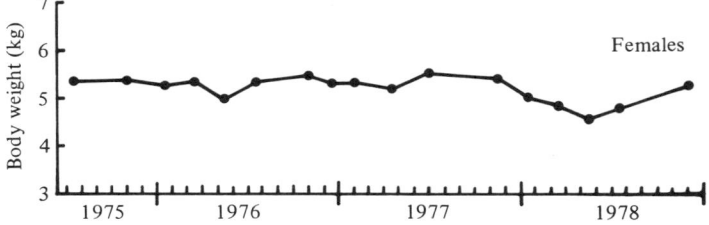

190   6 *Diet and nutrition of kangaroos and wallabies*

presented radiotracking data which indicated that the Tammars on Kangaroo Island expanded the size of their home ranges in summer, presumably because they were searching for better quality food as the annual grasses dried off. In this regard Andrewartha & Barker (1969) observed that Tammars foraged beneath *Acacia retinodes* trees during summer, suggesting that the animals may have been obtaining nitrogen from the seeds of this leguminous species.

Thus it is more likely that the Kangaroo Island Tammar suffers from an inadequate energy intake over the late summer and early winter. The greater output of faecal dry matter at this time of year (Barker, 1971) suggests that digestibility of the Tammar's diet is low during the summer months. Greater energy expenditure, as suggested by an increase in the size of their home ranges in this season, would tend to exacerbate any energy deficiency. There was no evidence of severe dehydration in Inns's (1980) study.

Since nitrogen and water do not appear to be limiting factors on Kangaroo Island with its higher annual rainfall and longer growing season, Tammar Wallabies from this island may be expected to be less well-adapted

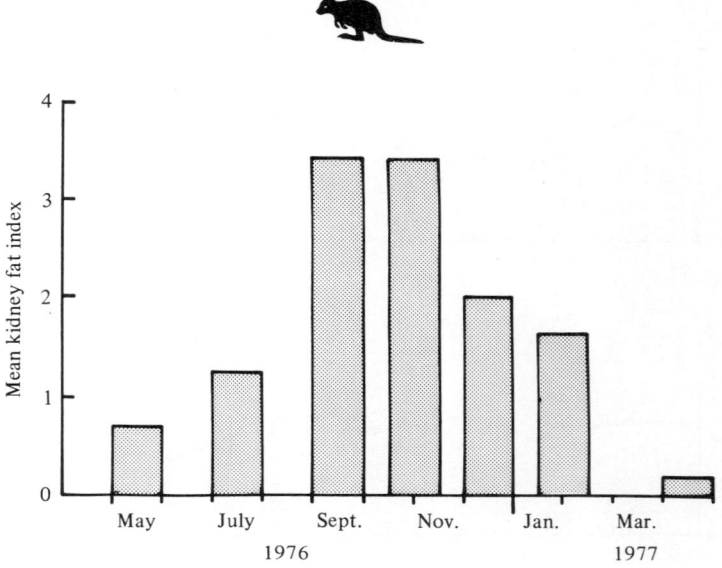

Fig. 6.10. Seasonal changes in the amount of fat present within the gut cavity of male *Macropus eugenii* on Kangaroo Island. The kidney fat index was based on a visual assessment of fat present in the gut cavity (kidney and mesenteric fat). After Inns (1980).

Nutrition and ecology 191

metabolically than Abrolhos Tammars to shortages of nitrogen and water. No direct comparisons between Abrolhos and Kangaroo Island Tammars have been reported, but they would be interesting in view of the possible physiological differences between Abrolhos and Garden Island Tammars already alluded to. Genetically the two Western Australian populations are identical (Richardson & McDermid, 1978), but the Kangaroo Island population differs in the locus of the pyruvate kinase allele. Nevertheless the genetic similarity coefficient (96%) is still high enough not to warrant any formal separation (Richardson & McDermid, 1978). The degree of physiological similarity is not known.

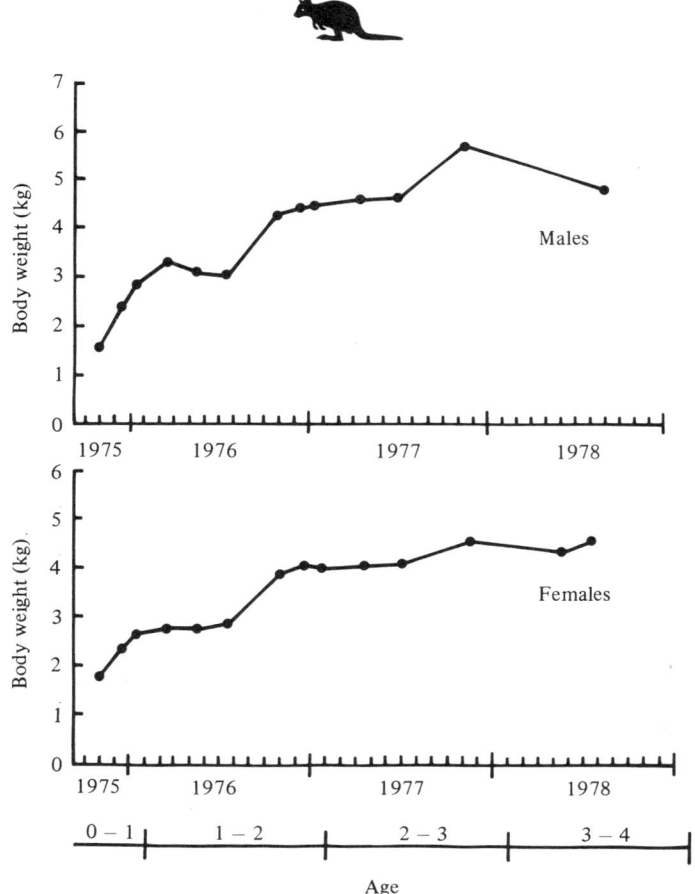

Fig. 6.11. Growth pattern of young male and female *Macropus eugenii* on Kangaroo Island. After Inns (1980).

192   6 Diet and nutrition of kangaroos and wallabies

Forest species – the grey kangaroos and pademelons
Grey kangaroos

In general grey kangaroos occur in areas of higher, more predictable rainfall than the arid-zone kangaroos. Their preferred habitat includes woodlands, shrublands, open forests and semi-arid mallee open scrubs (Russell, 1974). Two distinct species are recognised by Kirsch (1977a, c), *Macropus giganteus* (the Eastern Grey Kangaroo) and *M. fuliginosus* (the Western Grey Kangaroo). Their distributions overlap in the south-west of New South Wales (Fig. 6.12).

The Eastern Grey Kangaroo. The Eastern Grey Kangaroo is common throughout its range. Nevertheless relatively few studies have been made of its field biology, apart from movements and feeding patterns. The reports of Caughley (1964), Kirkpatrick (1965) and Bell (1973) all contain the remark that *M. giganteus* spends most of the day resting in shady open forest or woodland close to more open feeding areas. Partial clearing of woodland improves the habitat of *M. giganteus* by increasing the supply of grass (Calaby, 1966).

The most comprehensive physiological studies on the nutrition of the

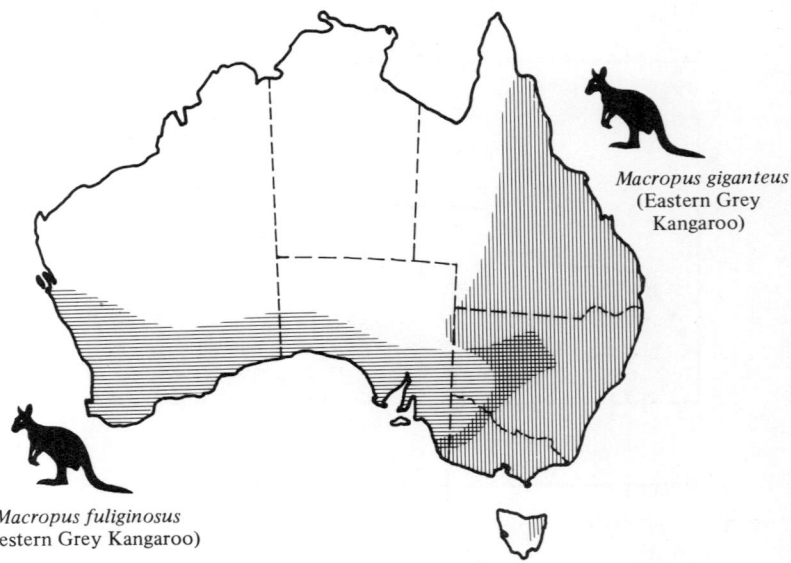

Fig. 6.12. Distribution of *Macropus fuliginosus* (the Western Grey Kangaroo) (horizonal lines) and *Macropus giganteus* (the Eastern Grey Kangaroo) (vertical lines). After Poole (1979).

Eastern Grey Kangaroo have been carried out at the University of New England. Foley, Hume & Taylor (1980) established the maintenance nitrogen requirement of *M. giganteus* fed a series of diets based on chopped oaten hay with graded levels of casein. The value of 270 mg truly digestible nitrogen$\cdot$kg$^{-0.75}\cdot$day^{-1} is close to the maintenance requirements of other macropodines included in Table 1.7, with the exception of the Euro which has the lowest requirement so far recorded for a macropodine (Brown & Main, 1967), and the Red-necked Pademelon which has the highest (Hume, 1977b). The maintenance requirement of the sympatric Wallaroo (*M. robustus robustus*), 240 mg truly digestible nitrogen$\cdot$kg$^{-0.75}\cdot$day^{-1}, was similar to that of *M. giganteus* (and thus significantly higher than that of the other subspecies of *M. robustus*, the arid-zone Euro). Analysis of the stomach contents of *M. giganteus* and *M. robustus robustus* showed that there were no significant differences between the two species in the nitrogen or fibre content of feed consumed in summer or winter. From the results of their feeding experiments in the laboratory Foley *et al.* (1980) concluded that nitrogen intakes were probably adequate to maintain both species in positive nitrogen balance throughout the year in the two study areas, one of which consisted largely of improved pasture species, but the other only of native pasture species.

Hume & Dellow (1980b) estimated the rate of production of volatile fatty acids (VFA) *in vitro* in *M. giganteus* shot in the field while grazing either native or improved pasture in late winter. Results were comparable with or slightly lower than estimates made by the same technique in the laboratory with animals maintained on a chopped lucerne hay diet. Again in line with laboratory results VFA production in the field animals was slower (mean of 20.5 μmol$\cdot$ml$^{-1}\cdot$h^{-1}) in the tubiform forestomach than in the sacciform forestomach (29.3$\pm$3.6 μmol$\cdot$ml$^{-1}\cdot$h^{-1}), and declined along the length of the tubiform region. VFA production in the hindgut was slower still, 10.5$\pm$1.0 μmol$\cdot$ml$^{-1}\cdot$h^{-1}.

The Western Grey Kangaroo. The Western Grey Kangaroo has received less attention. Prince (1976) conducted a series of physiological experiments comparing *M. fuliginosus* with the Euro and the Red Kangaroo in both the laboratory and the field. Fermentation rates were measured in the forestomach of animals shot in the field. The sampling procedure covered most of the 24 hours of the day, enabling estimates to be made of total daily VFA production in each species. There were two peaks of fermentation activity, corresponding to peaks of grazing activity in the early morning and late afternoon (Fig. 6.13).

Fermentation rates indicated that VFA production rate was higher and total daily VFA production greater in wild Western Grey Kangaroos than in either Euros or Red Kangaroos (Fig. 6.14). This could be a reflection of either a basic physiological difference between *M. fuliginosus* and the two arid-zone kangaroos, or a higher nutritive value of pasture grazed by *M. fuliginosus* in the south-west corner of Western Australia compared with that of arid-zone pastures. The higher nitrogen concentration in forestomach digesta of *M. fuliginosus* (Fig. 6.15) suggests that the Western Grey was consuming higher quality plant material than were the other two species. However, when all three species were maintained on common diets in the laboratory, dry matter digestibility of a low-nitrogen diet was highest in *M. fuliginosus*. This is good evidence that there is indeed a difference in digestive efficiency between the Western Grey Kangaroo and the Euro and Red Kangaroo.

Prince (1976) also found that total and inorganic phosphorus concentrations in forestomach digesta fluid of field animals were higher in *M. fuliginosus* than in the Red Kangaroo, although not the Euro. He suggested that this difference in phosphorus concentration could reflect differences between the Euros and Western Greys on the one hand and Red Kangaroos on the other in the recycling of endogenous phosphorus to the stomach via salivary secretions, and hence differences in their capacity to maintain conditions favourable for microbial fermentation. This suggestion

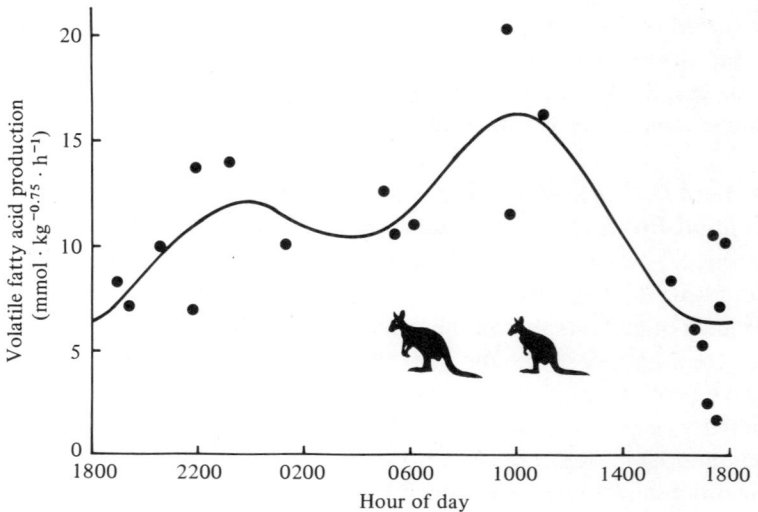

Fig. 6.13. The pattern of volatile fatty acid production in *Macropus fuliginosus*, *M. rufus* and *M. robustus erubescens* determined *in vitro*. The two peaks of production coincided with feeding activity in the evening and early morning. After Prince (1976).

is in general agreement with differences in fibre digestion and nitrogen and sulphur retention found by Hume (1974) between Euros and Red Kangaroos in the laboratory. However, because of the probable confounding effects in field animals of differences in quality of plant material grazed by each species, the differences found by Prince (1976) in phosphorus concentrations in the field would have to be shown to also hold in laboratory animals maintained on common diets before his suggestion of differences in phosphorus recycling could be accepted as fact.

Pademelons

The range of the genus *Thylogale* extends from the Bismarck Archipelago (3°S latitude), through New Guinea (*T. bruijni*) and the east coast of the Australian mainland (*T. stigmatica, T. thetis*), to Tasmania

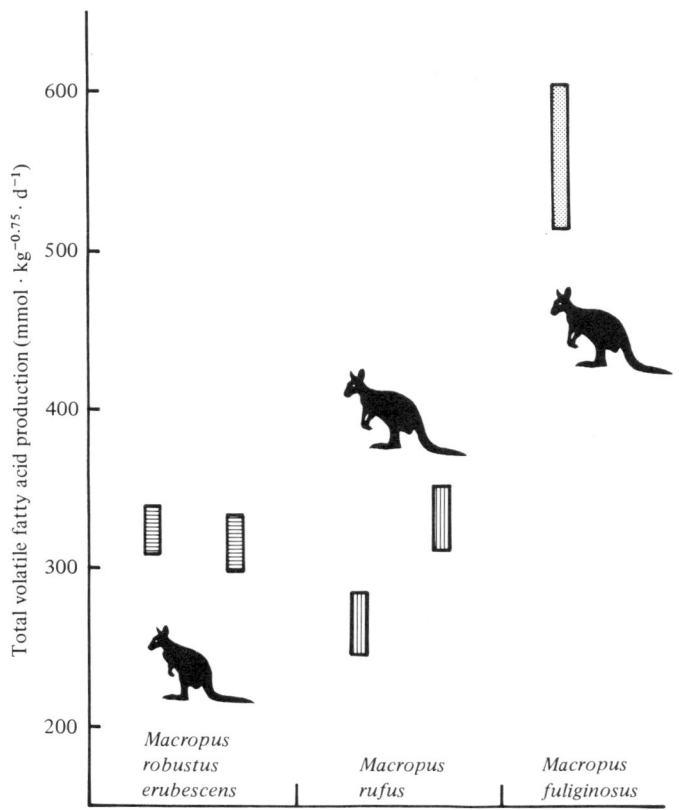

Fig. 6.14. Total daily volatile fatty acid production in *M. robustus erubescens*, *M. rufus* and *M. fuliginosus* measured in field animals by an *in vitro* incubation technique. After Prince (1976).

196   6 *Diet and nutrition of kangaroos and wallabies*

(*T. billardierii*) at 43°S latitude (Fig. 6.16). The habitats occupied by *Thylogale* have not been described in detail but the primary requirement appears to be dense forest vegetation, either rainforest (closed forest) or wet sclerophyll forest (tall open forest).

Pademelons are most numerous at the forest edge, from which they move short distances at night to feed in open areas (Calaby, 1966). Johnson (1977) found that radio-tracked *T. thetis* rarely moved more than 70 m from their rainforest refuge when feeding in open fields at night. During the day they moved up to 500 m into the rainforest, apparently searching for food, since Johnson (1980) often saw them feeding on ground vegetation. Although the population of *T. thetis* studied by Johnson (1977) near Dorrigo in north-eastern New South Wales occupied habitat that had

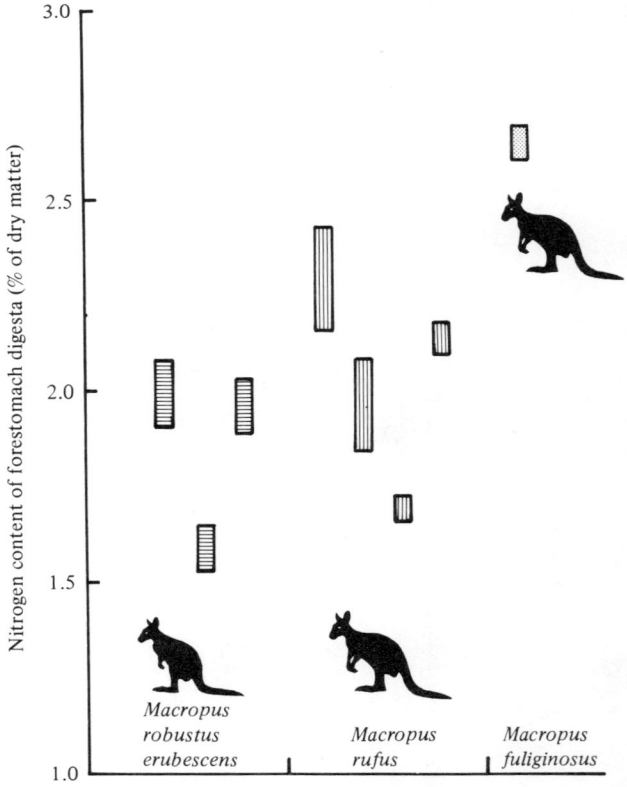

Fig. 6.15. The nitrogen content of forestomach digesta in *M. robustus erubescens*, *M. rufus* and *M. fuliginosus* shot in the field. The high nitrogen content in *M. fuliginosus* coincides with the high fermentation rate measured in that species. After Prince (1976).

Nutrition and ecology

been greatly modified by European man, Johnson (1980) argued that the present mosaic of open pasture and dense rainforest probably resembles the situation obtaining before the advent of European man when there were, even then, large, naturally occurring grasslands bordering rainforest, possibly as fire climaxes resulting from aboriginal activities (Calaby, 1966).

Thus the habitats of the four *Thylogale* species are quite different from

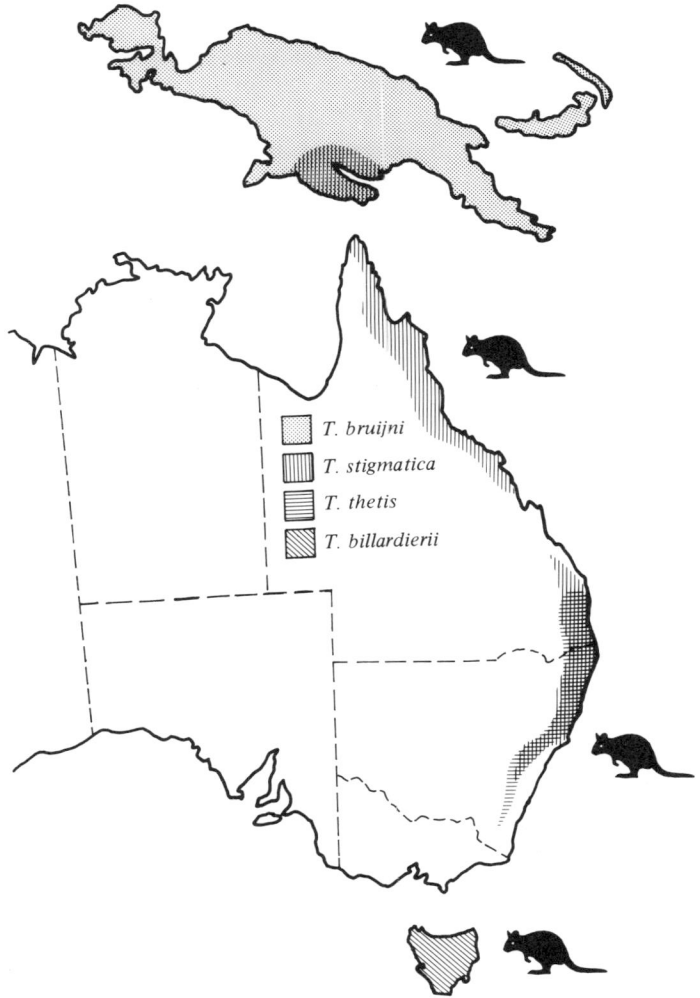

Fig. 6.16. Distribution of the four species of the genus *Thylogale*. The habitat of each species is similar, being moist forest, either wet sclerophyll forest or rainforest. After Johnson (1977).

those of the island wallabies *Setonix* and *Macropus eugenii*. Although there is a winter trough in food availability, Johnson (1977) could find little seasonal variation in adult weight in the *T. thetis* population he studied near Dorrigo. He also found that wild female pademelons bred continuously, rather than once a year as do Tammars and Rottnest Quokkas. Nevertheless, the frequency of births did peak in autumn and spring. These peaks were spaced by the length of the average pouch life (181 days) and it is probable that females raise two young each year. The peaks in breeding at Dorrigo occurred at times which enabled females to carry small pouch young and to be under little nutritional stress from lactation during the winter trough in food availability.

During his observations of feeding behaviour Johnson (1977) found that in their diurnal home range within the rainforest pademelons browsed on shrubs such as *Helichrysum diosmifolia* throughout the year, and grazed on herbs such as *Viola* sp. and on grasses. During winter months *Solanum mauritianum* (wild tobacco) was also eaten, despite its appreciable alkaloid content, common within the genus. Presumably the trough in food availability in winter forced the animals to expand their diet spectrum. In their nocturnal home range of open pasture area, grasses comprised the bulk of food items taken. In both diurnal and nocturnal sections of their home range *T. thetis* seemed to avoid eating dry vegetation whenever possible (Johnson, 1977).

These observations led Hume (1977a, b) to examine some nutritional and metabolic aspects of *T. thetis* in the laboratory. *T. thetis* is notable among the Macropodinae in having a maintenance nitrogen requirement at least twice that of any other species investigated so far (Chapter 1). Its inability to tolerate saline drinking water (Hume & Dunning, 1979) supports the contention that its kidneys are not adapted for conserving urea or water; *T. thetis* had a relative medullary thickness of 5.7, compared with the Tammar Wallaby of 7.2. Also consistent with these findings is the high water turnover rate reported by Dellow (1979) in the Red-necked Pademelon compared with the Tammar Wallaby. All these data suggest an animal ill-adapted to an environment where food quality or quantity may be seasonally limiting. Whether these features of *T. thetis* are shared by other members of the genus *Thylogale*, and by other macropodines restricted in their distribution to moist forests, such as the Parma Wallaby (*Macropus parma*), awaits investigation.

Summary and conclusions

The constraints imposed on field populations of several species of kangaroos and wallabies by shortages of energy, protein and water have been considered. This has provided the opportunity to extend the laboratory findings discussed in Chapters 1 and 5 to the animal in its natural habitat. Most of the field investigations have involved arid-zone species. The comparison between the Euro and the Red Kangaroo illustrated two different ways, one largely physiological, one largely behavioural, in which an animal can adapt to heat and frequent shortages of food and water during extended periods of drought.

The effect of the regular summer drought on two small wallabies, the Rottnest Island Quokka and the Abrolhos Tammar Wallaby, has also been the subject of detailed investigation. The finding that Quokkas occupying a more mesic habitat on the mainland close to Rottnest Island breed continuously throughout the year, while those on Rottnest breed only once, underscores the central role of food supply in the reproductive success of this species.

At present we know little about the field nutrition of many forest-dwelling macropodines. The Red-necked Pademelon has an unusually high maintenance requirement for protein and seems profligate in its usage of water, but are these features purely a phylogenetic trait or are they shared by other macropodine species from mesic environments?

The interpretation of laboratory findings with many species, not only those from rainforests, is limited by our incomplete knowledge of the animals' food habits and nutrient requirements in the wild. Recent and current attempts to estimate field energy and water budgets of marsupials from different environments should result in valuable additions to our understanding of the relationship between animal and environment. Our knowledge of the dietary habits of macropodines is continually, if slowly, expanding, and the relationship between diet and dentition among the Macropodinae has recently been re-examined. More will be said about this in the next chapter when the possible evolution of the Macropodidae is considered.

7

Herbivorous marsupials – the rat-kangaroos

The rat-kangaroos (subfamily Potoroinae) are the smallest members of the family Macropodidae. Often regarded as the more primitive of the macropodids, they nevertheless exhibit a number of highly specialised features. Their small size is perhaps the main reason why so many potoroine species are now either extinct or threatened with extinction. For instance, *Bettongia lesueur* is now found only on Barrow, Boodie, Bernier and Dorre Islands off the Western Australian coast. Its disappearance from the mainland where previously it was widely distributed can be directly linked to overgrazing by domestic stock of tussock grasses under which *B. lesueur* established burrows for refuge, and to competition by introduced rabbits for succulent food (Wood Jones, 1924). Predation by introduced foxes and feral cats also took its toll of this and other potoroine species.

Food habits

Rat-kangaroos from a variety of habitats are distinguished as a group in their food habits. Non-fibrous plant material appears to constitute the bulk of the diet of *B. penicillata* in south-western Australia (Kinnear *et al.*, 1979), of *B. lesueur* on Bernier and Dorre Islands (Ride & Tyndale-Biscoe, 1962), of *Potorous tridactylus* in Tasmania (Guiler, 1971), of *Aepyprymnus rufescens* in north-eastern New South Wales (F. Schlager, unpublished) and *Hypsiprymnodon moschatus* in north Queensland.

Kinnear *et al.* (1979) could find no evidence that *B. penicillata* in a forested area in the south-west of Australia ate fibrous plant material. The diet consisted principally of the sporocarps of hypogeous fungi, although gum exudates produced by shrubs and trees were also eaten. However, the fungi appeared to represent the major source of protein for the animals. Although the fungal sporocarps are protected by an indigestible outer cover, this cover is removed by *B. penicillata*, and the inner 'kernel' along with some adhering spores is ingested. Analysis of the kernel showed that it contained 8–10% crude protein, 42% lipid, and less than 1% ash. As a protein source the kernel appeared to be deficient in lysine but very high in methionine, indicating a grossly imbalanced protein (Kinnear *et al.*, 1979).

Studies on the fire ecology of *B. penicillata* by Christensen (1977) revealed that after bushfires *B. penicillata* survived and returned to feed on the fungi, which, being hypogeous, escaped incineration. The animals were able to maintain weight on this material, the only food source left, while Tammar Wallabies (*Macropus eugenii*) which share the same habitat lost weight after fire since their principal food source of grass had disappeared.

Sampson (1971) found a much broader food spectrum in his study of *B. penicillata* in a fauna reserve in the wheatbelt of Western Australia. Scats contained the remains of both mono- and dicotyledonous leaf material, testa of the seeds of a number of plant species, roots and tubers, bark, and cuticle from several species of arachnids and insects. Although quantitative results were not given there was no mention of fungi in the diet. Sampson (1971) felt that quantitative evaluation of the diet of *B. penicillata* was not justified because of their feeding behaviour. In the case of seeds, they chewed the seed and spat out the epidermal material and hard parts, swallowing only the soft, easily digested contents. When insects were eaten the animals cracked the cuticle and sucked out all the soft parts of the body, swallowing only after spitting out the hard parts. This would be expected to lead to an underestimate of the contribution of seeds and insects to the diet of *B. penicillata*.

Another aspect of the feeding behaviour of *B. penicillata* observed by Sampson (1971) was the burying of seeds, not only of wheat but also the very hard nut of the quandong (*Santalum acuminatum*). He suggested that *B. penicillata* may dig up the seeds and eat the soft parts once germination had begun. This behaviour could also be an important factor in the dispersal of some plants such as the quandong which have very hard seed coats.

On Bernier and Dorre Islands in Shark Bay, Western Australia, Ride & Tyndale-Biscoe (1962) found that rectal pellets collected from *B. lesueur* contained no recognisable epidermis. There were some small brown particles which may have been seed coats but most of the tissue was non-epidermal and possibly the remains of roots. Scratchings of *B. lesueur* were often found around the base of the succulent *Carpobrotus equiliaterale*.

Finlayson (1958) noted that in central Australia the roots of the procumbent *Boerhaavia diffusa* formed an important part of the diet of *B. lesueur*, and that bulbs and tubers were also generally eaten.

The diet of the Long-nosed Potoroo (*Potorous tridactylus*) has been examined in Tasmania by Guiler (1971) and its habitat requirements by Heinsohn (1968). Like *Bettongia* species, potoroos dig shallow holes when seeking food, and the manus has median claws that are long, strong and well adapted for digging. Microscopic examination of faecal pellets suggested that *P. tridactylus* was an omnivorous species but largely dependent upon plant material for much of the year (Table 7.1). Insects formed only 1–2% of the diet in winter, but this increased to 21% in summer as insect abundance increased. However, the most important foods during all months of sampling were fungi. Other minor items occurring regularly in the faeces were bryophytes and sedges. Cranberries (*Astraloma humifusum*) were also eaten except in winter when berries were absent from the plants. Grasses were present in the faeces only during winter, presumably when total food availability was lowest.

Plate 7.1. A juvenile *Aepyprymnus rufescens* (Rufous Bettong or Rufous Rat-kangaroo). (Ray Williams.)

Guiler (1971) realised that scat analysis may over-emphasise the importance of fungi in the diet of *P. tridactylus*, since only the hyphae were digested, the fruiting bodies appearing in the faeces. In contrast, other plants and the insects had been extensively digested. However, examination of stomach contents showed that the proportions of the various food items were substantially the same as those found in the scats.

The wide dietary spectrum and digging habit of *P. tridactylus* are strong survival factors. This was evident after bush fires in Guiler's (1971) study area when potoroos survived although most other mammals perished. The potoroos' survival was in large part due to the continued availability of food below ground, as shown for *B. penicillata* in Western Australia by Christensen (1977). Guiler (1971) also thought that the large home range of *P. tridactylus* was associated with its utilisation of fungi as its main food source. In order to obtain sufficient food from this source potoroos have to use a number of widely scattered feeding areas.

Aepyprymnus rufescens (the Rufous Bettong or Rufous Rat-kangaroo) (see Plate 7.1) is being studied in north-eastern New South Wales. Calaby (1966) and Kaufmann (1974) described its preferred habitat as being tall woodland with a ground cover of tussock grasses. It shelters during the day in nests made of thin-stemmed grasses, particularly *Poa*, and located in or under grass tussocks, at the bases of trees or against logs, and under fallen branches or shrubs. Bundles of grass are carried with the tail. Calaby (1966) often saw *A. rufescens* grazing on improved pasture even when this was closely grazed by cattle, and stomachs of specimens collected contained

Table 7.1 *Monthly faecal pellet analysis from 31* Potorous tridactylus *in Tasmania; the amount of food is expressed as percentages by area on microscopic slides*

	May	June	July	Aug.	Sept.	Oct.	Nov.	Dec.
Insects	0	2	2	1	0	9	14	21
Grasses	8	7	8	2	4	0	0	0
Sedges	0	1	1	1	3	7	1	1
Forbs	6	4	0	0	0	5	0	0
Seeds	0	1	0	0	2	1	1	3
Bryophytes	2	0	2	0	0	1	0	0
Unidentifed plant	6	4	4	4	9	5	4	8
Fungi	78	83	83	92	82	71	81	67

After Guiler (1971).

much introduced clover and the grass *Paspalum dilatatum*. Schlager (unpublished) has recently carried out a field study of *A. rufescens*, and has often observed them at night digging up and eating the swollen tap root of the introduced flat weed *Hypochaerus radicata* and thistles in open fields. By day the diggings are easily identified by the discarded tops of these weeds. Other items seen being eaten by Schlager were hypogeous fungi associated with *Casuarina* roots and gum exudate at the base of *Acacia* trees. Thus it seems that *A. rufescens*, although observed grazing by Calaby (1966), probably selects non-fibrous plant material whenever possible.

Hypsiprymnodon moschatus is found only in tropical rainforest along the coast and slopes of the ranges in north Queensland (Poole, 1979). Hypogeous fungi appear to be a major food item, since the animals are frequently observed digging in the litter on the floor of the rainforest, even by day. The importance of invertebrate animal material in the diet of *H. moschatus* is unknown, but in its rainforest habitat this could be expected to be considerable.

The digestive tract

The simplest potoroine stomach morphology is seen in *H. moschatus*. It has been described by Carlsson (1915) and by Heighway (1939) (Fig. 7.1).

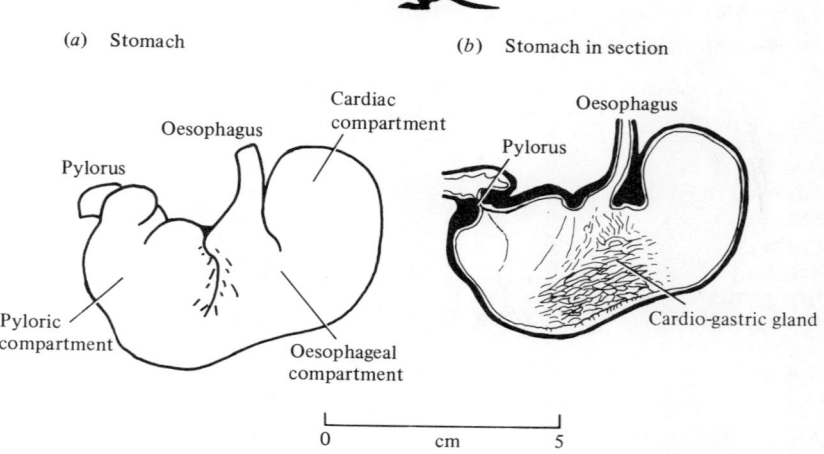

Fig. 7.1. The stomach of *Hypsiprymnodon moschatus* (a) external features, (b) in longitudinal section. After Heighway (1939).

The digestive tract

Carlsson (1915) considered the *Hypsiprymnodon* stomach to be simple and sacciform, and more similar to that of *Trichosurus* than to any of the other Potoroinae. The simple stomach, together with other features of *H. moschatus* such as its pronounced plagiaulacoid premolars has led to the suggestion by some that it should be placed in a third sub-family within the Macropodidae (Kirsch & Calaby, 1977). However, although externally there is no evidence of haustration as seen in the macropodine stomach, Heighway (1939) referred to two fairly distinct transverse grooves on the

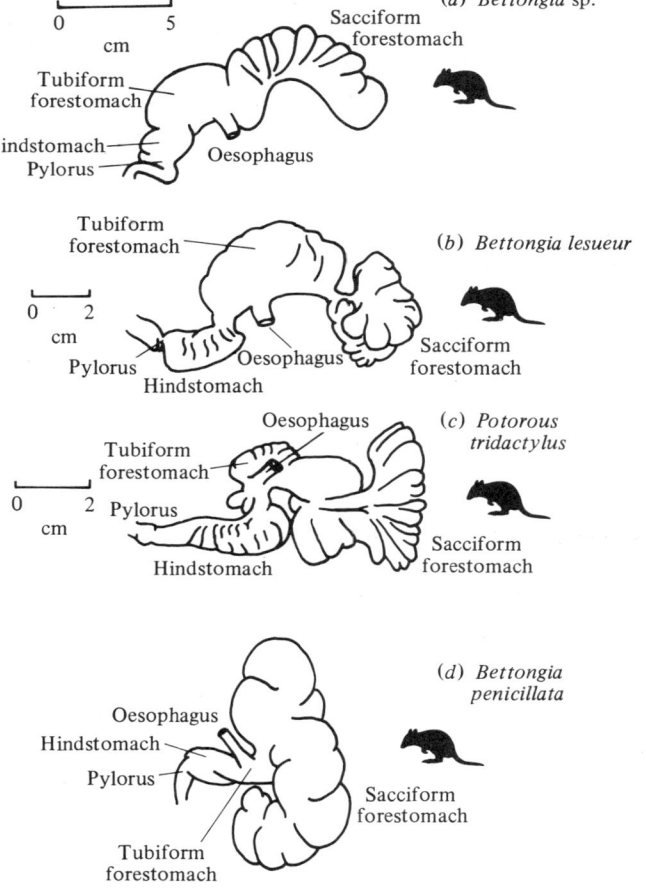

Fig. 7.2. The stomachs of (*a*) *Bettongia* sp. (after Schultz, 1976), (*b*) *Bettongia lesueur* (after Langer, 1980a), (*c*) *Potorous tridactylus* (after Langer, 1980a) and (*d*) *Bettongia penicillata* (after Kinnear *et al.* 1979). In each case the sacciform forestomach is the largest gastric region.

Hypsiprymnodon stomach, one on either side of the oesophageal opening. The grooves are most prominent on the lesser curvature, and disappear before they reach the greater curvature. She referred to three 'compartments' produced by the internal ridges associated with the grooves as the cardiac, oesophageal and pyloric compartments.

The gastric mucosa is entirely glandular. In addition, the mucosa of the oesophageal compartment is organised into a thick, heavily plicated structure, with numerous glandular crypts opening on and between the folds. From its structure and its relationship to the cardia, Heighway (1939) considered this to represent the cardio-gastric gland described in Chapter 4 for the wombat and Koala. She concluded that the *Hypsiprymnodon* stomach was more or less intermediate between the simple form found in the Phalangeridae and the complex type characteristic of the Macropodidae.

The total length of the small intestine in the specimen examined by Heighway (1939) was 90 cm. The interior of the duodenum was devoid of rugae, and the mucosa was closely villous. The remainder of the small intestine was villous and rugose, with a large number of small lymphoid nodules. The hindgut was well developed. The simple caecum was 5 cm in length and the total length of the colon was 57 cm. There was only limited haustration of the caecum and colon, the haustra being maintained by taeniae which were almost indiscernible. The internal surface of the caecal wall was only slightly rugose, but that of the colon markedly so. Heighway (1939) also remarked on the ileo-colic junction which projected prominently on the medial wall of the caecum, suggesting an efficient valvular action.

The stomach of all three of the genera *Potorous*, *Bettongia* and *Aepyprymnus* is more complex than that of *Hypsiprymnodon*, and more like the macropodine stomach discussed in Chapter 5. The stomachs of the three former genera are shown in Fig. 7.2.

The cardiac compartment of Heighway (1939) appears to be greatly expanded into a cul-de-sac (Kinnear *et al.*, 1979) or sacciform forestomach region (Langer, 1980a). The wall of this region is differentiated into taeniae, haustra and semi-lunar folds. The oesophageal compartment of Heighway (1939) is then equivalent to the tubiform forestomach region of the Macropodinae (Dellow, 1979). The wall of this region is largely undifferentiated (Langer, 1980a) and there is no gastric sulcus in *Bettongia* or *Aepyprymnus*. Although a gastric sulcus is present in *Potorous* (Langer, 1980b), it is poorly developed. Heighway's (1939) pyloric compartment corresponds to the term hindstomach, since it is the region of acid secretion (Kinnear *et al.*, 1979).

Thus it is possible to assign similar nomenclature to both the potoroine and macropodine stomachs. There are, however, several significant differences. First, the potoroine sacciform forestomach is by far the largest gastric region, constituting perhaps as much as 75% of total stomach capacity. This contrasts with most macropodines in which the tubiform forestomach is the main region. Owen (1868) was the first to remark on this difference: 'The left or cardiac division is enormously developed; in relative proportion, indeed, it is surpassed only by the true ruminant stomachs, in which both the rumen and reticulum are expansions of the corresponding or cardiac moiety of the stomach.' The second major difference is that in the potoroines the oesophagus opens into the distal tubiform forestomach rather than near the sacciform–tubiform forestomach border (Langer, 1980a). Again, Owen (1868) was the first to comment on this: 'The relation of the stomach of a Potoroo to that of a Kangaroo may be concisely expressed by stating that the termination of the oesophagus in the former is removed from the commencement, or left, of the middle sacculated compartment to its termination.' Thus the distance between cardia and hindstomach is small. This may explain why there is no gastric sulcus in most Potoroines, which again contrasts with the majority of macropodine species.

A further difference lies in the fact that in the potoroine stomach there are no areas of squamous epithelium, the whole of the sacciform and tubiform regions of the forestomach being lined with cardiac glandular mucosa. In most macropodines squamous epithelium lines at least part of the sacciform forestomach as well as the gastric sulcus.

How important these differences are nutritionally must await further research on the function of the potoroine stomach. The only functional study to date has been that of Kinnear *et al.* (1979). These workers examined the composition of stomach contents, blood plasma and urine in eight *B. penicillata* captured in forested country in south-west Western Australia. The pH of the sacciform forestomach contents was about 5.8, and that of the tubiform region 5.0, suggesting that these were areas of active fermentation. This was confirmed by the presence of volatile fatty acids (VFA) at levels which were similar to those found by Kempton *et al.* (1976) in *Macropus giganteus* maintained on chopped lucerne hay, although lower than in *M. giganteus* shot while grazing in the field (Dellow, 1979). The pH of the hindstomach contents of *B. penicillata* was 2.8 (Kinnear *et al.*, 1979). Plasma urea concentrations were similar to those in *M. giganteus* fed chopped lucerne hay (Dellow, 1979), and the urinary ratio of urea nitrogen to total nitrogen was uniformly high (mean UR ratio of 0.83), indicating that either the diet selected by *B. penicillata* was high in protein, or that urea recycling to the gut of *B. penicillata* was unlikely to be of much significance to the nitrogen economy of the animal.

Possible evolution of the macropodid digestive system

Although much more information on the form and function of a wider range of macropodid species, especially the smaller browsing forms, is needed before we can fully understand their interrelationships, it is interesting to speculate on the possible evolution of their digestive systems. This has been done from both a functional (Hume, 1978) and an anatomical approach (Langer, 1980b).

Hume (1978) suggested that the large sacciform forestomach region of the Potoroinae may have evolved at least partly as a storage organ as well as a fermentation chamber. Predator evasion by these small macropodids depends on use of forest or other dense vegetation as refuge areas (Heinsohn, 1968). Feeding areas are generally in more exposed areas. It would be advantageous for potoroines to feed actively over short periods of time, but this strategy would depend on a storage area for ingested food.

According to the palaeontological evidence of Bartholomai (1972), the Macropodidae were represented in the early Oligocene by the sub-family Potoroinae, including a species of the extant genus *Bettongia*. Sanson (1976) considered all the Potoroinae to have a primitive dentition adapted to non-abrasive material (see Chapter 6). This is consistent with the known dietary habits of extant potoroines as discussed earlier in this chapter. The sub-family Macropodinae was almost certainly distinct from the Potoroinae by the end of the Oligocene, and before the appearance of grasses in the fossil record (Sanson, 1976). On this basis it might be expected that early macropodines would have been browsers, and would exhibit some features of the Potoroinae. The stomachs of the extant macropodines *Thylogale thetis* and *T. stigmatica* resemble the potoroine stomach in several ways: the sacciform forestomach is the largest gastric region, the tubiform forestomach is comparatively short, and there is no gastric sulcus. However, it is often unwise to assume similar structures across all members of a genus. In this instance *T. billardierii* does exhibit a gastric sulcus (Dellow, 1979). Nevertheless the dentition of all *Thylogale* species is of the browsing type (Sanson, 1976).

Grazing macropodine forms appeared a little later, in the Miocene, after the appearance of grasses. The stomach of extant grazing macropodines features a comparatively reduced sacciform forestomach, and instead enlargement of the tubiform forestomach into a coloniform structure (Owen, 1868; Moir *et al.*, 1956), as exemplified by *Macropus giganteus*

(Fig. 5.8). Hume (1978) considered these to be features which may well have evolved in response to the appearance of grasses in the early Miocene and their subsequence radiation since that time as world climates became progressively more arid.

Why should a tubiform stomach be more advantageous than a sacciform stomach to the large grazing macropodine species? To answer this question we must make another comparison, this time between the Equidae and the Ruminantia. There is evidence from both the laboratory and the field that, provided food availability is not limiting, horses are able to maintain their intake of grass at a higher level as the fibre content increases than can ruminants (Janis, 1976).

The reason for this difference appears to be anatomical. In the horse, in which fibre digestion occurs mainly within the haustrated colon and caecum, there is no structure comparable to the omasum and the reticulo-omasal orifice, either in the stomach or the hindgut, to impede the movement of coarse particles. As a result the intake of higher fibre diets by horses is not necessarily depressed. Darlington & Hershberger (1968) reported that horses maintained their intake of *Phleum pratense* (timothy) and *Dactylis glomerata* (orchard grass) as the grass matured. In contrast, dry matter intake by ruminants is usually depressed as fibre content increases. This is because coarse feed particles must be broken down to a certain size before they will leave the reticulo-rumen. On high fibre diets the rate of particle breakdown in the rumen is so slow that rate of passage out of the reticulo-rumen is depressed, rumen fill increases, and feed intake is inhibited; the end result is that the animal may not be able to satisfy its energy requirements even though food availability is not a problem (McClymont, 1968).

Thus Bell (1969) and Owaga (1975) have both reported that zebra selected plant species and plant parts of higher fibre content than did ruminants grazing the same low-quality herbage, despite the presumed lower digestibility of the more fibrous material.

The hypothesis developed by Hume (1978) is that the grazing macropodines evolved to utilise the large area of grassland in inland Australia not by the development of a rumen-like stomach, but instead a stomach more like a colon. This may have enabled them to utilise grasses which were often of high fibre content, not by retaining it in the forestomach longer to increase fibre digestion, but by increasing the rate of passage through the stomach at the expense of high fibre digestion, thereby maintaining their food intake and digestible nutrient supply at a higher level than would otherwise be possible.

The data of Foot & Romberg (1965), who compared the intake and digestion of roughages by young *Macropus rufus* (Red Kangaroos) and mature Corriedale wethers, can be used to support this argument (Table 7.2).

The young Red Kangaroos consumed less of the high-quality chopped lucerne hay diet than did the sheep. However, on the poor quality chopped oaten straw diet, although dry matter intake by both herbivores declined, intake by the Red Kangaroos was 40% greater than by the sheep. Apparent digestibility was lower in the Red Kangaroos on both diets, but the intake of apparently digestible dry matter by the Red Kangaroos on the high fibre diet was greater than by the sheep. Thus it appears from this study that the Red Kangaroo may be better able to maintain its intake of a high fibre diet than is the sheep. The reason why this difference between grazing macropodine species and sheep is not apparent in similar studies by McIntosh (1966), Forbes & Tribe (1970) and Hume (1974) is thought to be due to the fact that the latter studies involved only mature animals. As we saw in Chapter 1 the maintenance energy requirement of the Macropodinae is at least 20% below that of the sheep (Hume, 1974). Only if the total energy requirement of the macropodine was greater than that of the ruminant, as it was in the study of Foot & Romberg (1965), would a higher digestible energy intake be expected.

Similar studies with browsing members of the Macropodinae and with

Table 7.2. *Intake and digestion by three young Red Kangaroos and three mature sheep*

	Red Kangaroo	Sheep	Red Kangaroo compared with sheep (%)
Body weight (kg)	11.4	38.8	
Dry matter intake ($g \cdot kg^{-0.75} \cdot d^{-1}$)			
Lucerne hay	58.1	71.7	81
Oaten straw	40.4	28.9	140
Digestibility (%)			
Lucerne hay	54.0	62.9	86
Oaten straw	35.9	39.4	91
Digestible dry matter intake ($g \cdot kg^{-0.75} \cdot d^{-1}$)			
Lucerne hay	31.4	45.1	70
Oaten straw	14.5	11.4	127

After Foot & Romberg (1965).

the Potoroinae would be necessary to further test this idea. Whether the grazing Macropodinae were derived from browsing forms, which in turn were derived from the early Potoroinae, as suggested by Hume (1978), or whether the potorine and macropodine stomachs evolved independently of each other as suggested by Langer (1980b) is not known.

Conclusions

Because of our very incomplete knowledge of the Potoroinae little can be said about the adaptive significance of many features of the digestive system of this supposedly primitive macropodid group. One member of the group, *Hypsiprymnodon*, has a stomach often regarded as more closely allied to the Phalangeridae than to the rest of the Macropodidae, and there is debate as to whether this monotypic genus should be placed in a sub-family of its own. The stomach of other members of the Potoroinae is much more complex. Although food storage as part of a predator evasion strategy has been suggested as one function of the extraordinarily large sacciform region of their forestomach, the extent to which potoroines rely upon microbial digestion in this region for their nutrient supply is unknown. Although several species are known to be highly specialised in their feeding habits in that hypogeous fungi form a major part of their diet, it would be interesting to know the extent to which these animals can utilise higher fibre content foods such as grasses during seasons of the year when the availability of fungi and other more highly digestible foods may be low. Much could also be learned from more detailed studies of stomach form and function in the Potoroinae.

8

Mineral and vitamin nutrition of marsupials

Little mention has been made thus far of the role of minerals and vitamins in the nutrition of marsupials. This was an area of active research in the decade 1955–65, but little has been done since. The notable exception to this among the minerals is the metabolism of sodium and potassium, which has been frequently studied in relation to kidney function and water and electrolyte balance. The notable exception among the vitamins is ascorbic acid (vitamin C), which has now been investigated in some detail by Birney, Jenness & Hume (1979, 1980).

Most of the mineral studies have been conducted with the Macropodidae. There is little ecological justification for studying the mineral nutrition of carnivores, since the mineral composition of their principal food sources remains quite constant throughout the year. More likely to vary is the availability of their prey, but a shortage of prey would be expected to lead to a general energy deficit long before any problem of mineral deficiency was encountered. Omnivores are subjected to greater fluctuations than carnivores in mineral status, since plants generally vary in mineral composition throughout the year. However, the greater part of an omnivore's requirement for minerals is likely to be met from animal sources, not plant products such as nectar or seeds. This would tend to buffer omnivorous species from wide fluctuations in the availability and mineral composition of plant materials.

It is among the herbivores that the widest fluctuations in mineral status are encountered, particularly among grazers. Browsers may be restricted to foliage of low mineral content, but it is likely to be relatively constant throughout the year unless the preferred tree species are deciduous. Eucalypt foliage appears to be quite constant in its concentration of minerals, at least those elements analysed by Ullrey *et al.* (1981): calcium, phosphorus, sodium, potassium, magnesium, iron, copper, selenium and zinc, with no significant differences between young and mature leaves. However, Ullrey *et al.* (1981) also noted that the concentrations of phosphorus, sodium, zinc and selenium and possibly copper in *Eucalyptus* foliage either consumed or rejected by Koalas were often below the levels recommended for domestic sheep and horses. It must be concluded from this that the mineral requirements of Koalas, like their energy, protein and water requirements, are below those of most eutherians (Chapter 4). Presumably the same holds for other strictly folivorous marsupials such as the Greater Glider and the Ringtail Possums. In contrast to the situation faced by most browsers, grazers can be subjected to variations in mineral supply of some magnitude, depending on the soils and climate of their habitat. The limestone-derived sandy soils of Rottnest Island are poor in their status of most minerals. Thus early attempts to graze sheep on the island met with failure when the animals quickly developed a wasting disease now known to be a dual deficiency of copper and cobalt. This experience, together with the finding that many Quokkas on Rottnest were emaciated and anaemic at the end of the annual summer drought, led Barker (1960, 1961a, b) to embark on a study of the role of several minerals in the nutrition of the Rottnest Quokka.

Copper, molybdenum and sulphate

Grazing merino sheep require a minimum of 6 p.p.m. copper in dry pastures to avoid defective keratinisation of the wool (Underwood, 1977). Copper deficiency can be caused by a low intake of copper, or it can be induced on normal copper diets containing moderate levels of molybdenum and inorganic sulphate. Dick, Dewey & Gawthorne (1975) have proposed that there are three essential steps in the induction of copper deficiency in ruminants. These are (a) reduction of sulphate to sulphide in the rumen, (b) the reaction at relatively neutral pH of this sulphide with molybdate in the rumen to produce thiomolybdate, and (c) reaction of the thiomolybdate with the copper to give the very insoluble copper thiomolybdate, $CuMoS_4$.

It appears from the work of Barker (1960, 1961a) that the same three-way copper-molybdenum-sulphate interaction occurs in Quokkas. In an experiment with captive animals fed a diet containing 7.5 p.p.m. copper, drenching animals daily with 5 p.p.m. molybdenum and 1 g inorganic sulphate per 100 g dry matter intake significantly depressed blood and liver copper levels. However, even though the experiment continued for 20 weeks, copper deficiency anaemia was not induced.

Analysis of Rottnest plants showed that some species contained very little copper, moderate levels of molybdenum and high levels of inorganic sulphate (Barker, 1961b). Therefore blood samples were taken from Quokkas caught in the Lakes Area and at West End (Fig. 6.4) on nine collecting trips spaced over a period of 17 months (Barker, 1961a). Despite higher copper levels in favoured food plants such as *Sporobolus* from the Lakes Area than in plants from West End, animals from the Lakes Area had lower blood copper levels on all but three collecting trips. This was correlated with higher plant molybdenum levels in the Lakes Area, as demonstrated by higher blood molybdenum levels in Lakes Area Quokkas. This suggests that in this area blood copper levels were depressed by a higher molybdenum intake, as occurs in ruminants (Underwood, 1977).

In the Lakes Area blood copper levels were related to seasonal changes in the haemoglobin concentration in the blood. Quokkas from West End, on the other hand, had higher and more uniform blood copper levels throughout the year, despite wider fluctuations in haemoglobin concentration. It thus appears, both from the field observations and the laboratory experiment, that induced copper depletion alone does not cause the seasonal anaemia on Rottnest Island, but may be one factor associated with it in the Lakes Area at least (Barker, 1961a).

Two other indications came from this work. First, compared with the minimum copper requirements of merino sheep of 6 p.p.m. in dry feed, the Quokka requires only 3 p.p.m. copper (Barker, 1960), depending on molybdenum and inorganic sulphate levels in the feed. Second, female Quokkas in both study areas in spring had lower blood copper and haemoglobin levels than did males. At other times of the year, there was no sex difference. It appeared that the difference in spring blood copper levels could be attributed to the fact that in this season the females were suckling large pouch young. Copper levels in milk samples collected from lactating females in the spring of the following year ranged from 0.13–1.46 $\mu g \cdot ml^{-1}$. In all but 2 of the 25 lactating females sampled, the copper level in milk was higher than that in the blood of the same animal. This indicates that the Quokka, unlike the sheep and cow, but like the rat (Underwood, 1977), can concentrate copper in the milk. (Both the Quokka and the rat also concentrate iron in the milk, something which man and domestic ruminants cannot do (Kaldor & Ezekiel, 1962).) Barker (1962) considered that the maintenance of these milk copper levels could impose sufficient strain on the copper reserves of the females, after a prolonged period of lactation, to be responsible for the low levels of copper in the blood observed in the spring.

Cobalt and vitamin B_{12}

On the basis of evidence from ruminant studies, another possible factor in the emaciation and anaemia experienced by Quokkas on Rottnest was a deficiency of cobalt. Plant cobalt levels on Rottnest were approximately 0.03 p.p.m. (Barker, 1960), much lower than the cobalt requirements of sheep, viz. 0.08–0.11 p.p.m. (Underwood, 1977). Barker (1960) found that captive Quokkas were able to survive on diets containing as little as 0.01–0.03 p.p.m. of cobalt for at least 12 weeks. During this period liver cobalt levels fell from 0.3–0.7 p.p.m. to 0.08–0.19 p.p.m. (In sheep, liver cobalt levels fall from about 0.15 p.p.m. to 0.02 p.p.m. in the deficient state (Underwood, 1977).) However, no symptoms typical of cobalt-deficient sheep appeared in the Quokkas, and no anaemia developed. Although 12 weeks may not be long enough to deplete an animal's cobalt stores, the results nevertheless suggest that cobalt deficiency was not directly associated with the seasonal anaemia in the Rottnest Quokka (Barker, 1960).

In ruminants a cobalt deficiency is actually a vitamin B_{12} deficiency. The vitamin is synthesised by bacteria in the reticulo-rumen and in other parts of the gut. Thus on low-cobalt diets there are reductions in blood and liver not only in cobalt concentrations but in vitamin B_{12} levels as well. At the end of Barker's (1960) 12-week feeding period, vitamin B_{12} concentrations had fallen from 3–20 $\mu g \cdot ml^{-1}$ to 1–2 $\mu g \cdot ml^{-1}$ in plasma and from 0.8–1.5 $\mu g \cdot g^{-1}$ dry weight of liver to 0.4–0.8 $\mu g \cdot g^{-1}$. By way of comparison plasma B_{12} levels in Quokkas on Rottnest ranged between 0.5 and 3.0 $\mu g \cdot ml^{-1}$, being lower in juveniles than in adults, and lower in West End animals than in Lakes Area animals. This latter difference was not correlated with differences in plant cobalt levels (Barker, 1960). Nor were seasonal differences in plasma B_{12} concentrations correlated with anaemia, since the lowest plasma B_{12} levels were recorded in spring, not autumn. Thus the field observations supported Barker's (1960) conclusion from his laboratory experiment that cobalt deficiency, like copper deficiency, was probably not directly linked with seasonal anaemia in the Rottnest Quokka.

Selenium and vitamin E

For several years during studies on captive Quokkas from Rottnest Island, workers observed development of a paralysis of the hind limbs in animals maintained on a diet of high-protein (17–21% crude protein) commercial sheep pellets consisting of bran, pollard, oat meal, linseed meal, whale meal, molasses, urea and mineral salts. The disorder is characterised by weakness of the hind limbs which begins insidiously and progresses rapidly to complete paralysis. There is marked wasting of the muscles of the pelvic girdle and the disease invariably terminates in death (Kakulas, 1961). The same disorder has been observed in captive *Macropus eugenii* and *Thylogale thetis* maintained on lucerne hay by the author at Armidale. Histologically, there are marked degenerative changes in the muscles of the hind limbs, lesions typical of a deficiency of vitamin E (α-tocopherol) in the diet.

Kakulas (1961) demonstrated convincingly that the lesions could be completely reversed by oral dosing of affected Quokkas with 200–600 mg vitamin E daily for several days. In lambs, the condition (called 'white muscle disease' because of the pallor of the pelvic and femoral groups of muscles) can also be prevented by the addition of as little as 0.1 p.p.m. selenium to vitamin E-deficient diets. Thus in ruminants, and many other species, selenium has been found to be an effective substitute for vitamin E (Underwood, 1977). One species in which it is not is the rabbit. In further experiments Kakulas (1963a) demonstrated that supplementation of the commercial sheep pellets used before (Kakulas, 1961) with selenium (0.5 g per kg body weight) was without any prophylactic effect. We have found the same with *M. eugenii* maintained on lucerne hay. These results indicate that the Quokka (and the Tammar) belongs to the small group of animals in which nutritional muscular dystrophy is not prevented by trace amounts of selenium.

During his experiments Kakulas noticed that the smaller the size of enclosure, the higher was the incidence of nutritional muscular dystrophy in captive Quokkas. In a formal experiment with four animals in enclosures of three different sizes (1.2, 9.0 and 30 m^2) replicated four times, Kakulas (1963b) confirmed that the size of the enclosure was an important factor in the development in the Quokka of muscular dystrophy. Apparently the additional stress of crowding increases significantly the vitamin E requirement. In all cases, however, administration of the vitamin was completely effective in preventing paralysis in animals in similar and even smaller enclosures.

To ascertain whether the apparently high requirement for vitamin E by the Quokka was a factor in the seasonal debility of the Quokka on Rottnest Island, Kakulas (1966) measured the vitamin E status of plants and animals throughout the year. The vitamin E status of the animals was judged by the rate of haemolysis of collected blood samples, since this is thought to be directly affected by vitamin E (Kakulas, 1966). Results of these tests suggested that in spring the animals were receiving adequate vitamin E, but that during summer their vitamin E status was marginal. This correlated well with vitamin E levels in plants such as *Atriplex* sp. and *Sporobolus* sp. which were highest in winter and spring, but not with the results of muscle biopsy tests carried out concurrently on the same animals; the proportion of animals with muscle lesions was actually highest in winter. Also, any muscle lesions found were very mild. It would seem from these findings that vitamin E is not in itself an ecological problem for Quokkas on Rottnest Island.

Sodium and potassium

The ability of Tammar Wallabies to satisfy part of their water requirement by drinking sea water was discussed in Chapter 6 in relation to the capacity of their kidneys to excrete a concentrated urine. Also discussed in that chapter was the possibility that salt-loading of Quokkas on Rottnest Island contributes to their seasonal debility. This may be particularly so at West End where lack of fresh surface water leads to consumption of succulent halophytic vegetation (Storr, 1964a). Thus neither the Tammar nor the Rottnest Quokka lives in an environment in which sodium is ever likely to be in short supply. The generally high urinary sodium/potassium ratio in Red Kangaroos and Euros in western New South Wales was considered by Dawson & Denny (1969) to reflect the contribution of halophytic chenopods, principally *Atriplex* (saltbush) and *Kochia* (bluebush), to the diets of these two large kangaroos. The availability of these halophytes means that sodium availability is unlikely to present a problem to kangaroos in this region.

In contrast, at least two environments have been demonstrated to be sodium-deficient. One of these is the Snowy Mountains in south-eastern New South Wales (and to a lesser extent the grasslands surrounding Canberra nearby) where Blair-West *et al.* (1968) examined adaptations to sodium deficiency in the Eastern Grey Kangaroo (*Macropus giganteus*) and the Common Wombat (*Vombatus ursinus*). The other area is north-western Australia, the site of Ealey's (1967a, b) study of the Euro. Main (1970) concluded that both Euros and Red Kangaroos in this region were periodically affected by a sodium deficiency.

The environment of north-western Australia is just as arid as that of western New South Wales. The difference lies in the absence of halophytic chenopods in the vegetation of the north-west (Storr, 1968). As a consequence the sodium content of the diet of Euros and Red Kangaroos in this region is much lower than that of the populations in western New

South Wales studied by Dawson & Denny (1969). Lower mean urinary sodium/potassium ratios in both Euros (0.46 versus 0.89) and Red Kangaroos (0.43 versus 1.15) (Main, 1970 and Dawson & Denny, 1969) reflect the lower sodium intakes.

Main (1970) found that during a period of good rainfall and abundant plant growth in the north-west about 30% of Euros and Red Kangaroos had much lower body weights than the rest of the population. Further, these low body weights were correlated with very low urinary sodium excretion as indicated by very low sodium/potassium ratios in the bladder urine of shot animals. It was apparent that these animals were sodium depleted.

Main (1970) tested this idea in the Euros. By assuming that the daily excretion of creatinine by the Euro was constant per unit of body weight (Fraser & Kinnear, 1969), from the amount of creatinine in a known urine volume Main (1970) calculated that there were no gross differences in daily urine production between animals with average or with very low sodium concentrations in the urine. This confirmed that the low sodium concentration of the urine was not due to dilution resulting from high urine flows. The explanation for the very low urinary sodium excretion in some animals came from analysis of plant material in the mouth of Euros shot while grazing. These animals appeared to be eating mainly spinifex (*Triodia*) and other native grasses which grew in deep sands and showed no sodium. On the other hand, animals with average sodium excretion were found to be eating mainly the grasses *Cenchrus* and *Eriachne*, which grew on loams and clays between the sand dunes and showed sodium concentrations between 20 and 450 mEq·kg^{-1} dry matter. These levels are quite high compared with grasses from the sodium-deficient Snowy Mountains (Table 8.1). Thus the differences in urinary sodium excretion, at least in the Euros, were probably due to differences in the diet of this sedentary species. The low body weights of sodium-deficient Euros and Red Kangaroos were probably due to the fact that these animals would be unable to rehydrate fully by drinking since to do this would mean increased sodium losses in the increased urine flow. Low water intakes would depress intakes of dry matter (and therefore energy), leading to poor body condition, even though food was abundant.

Blair-West *et al.* (1968) examined the adaptations to a constant sodium deficiency in Snowy Mountain *M. giganteus* and *V. ursinus* by comparing a number of physiological parameters in animals from that environment with those in the same species near the coast in Victoria. The low sodium status of soils and grasses from the Snowy Mountains and Canberra are compared with soils and plants from western New South Wales and coastal Victoria in Table 8.1.

As can be seen from Table 8.2 sodium content of urine of both *M. giganteus* and *V. ursinus* in the Snowy Mountains was virtually zero. The value in coastal *M. giganteus* was similar to that reported by Dawson & Denny (1969) in the Euro during mild conditions in western New South

Table 8.1. *Sodium content of soils and grasses from four environments. Compared with coastal Victoria and western New South Wales, the Snowy Mountains and Canberra grasslands are severely sodium-deficient*

	Sodium concentration (mEq·kg^{-1} dry weight)			
	Snowy Mountains	Canberra	Western New South Wales	Coastal Victoria
Soil	0.5	0.8	3.6	3.3
Grass (mixed species)	0.2–8.5	5–10	96–200	68–309
Saltbush (*Atriplex*)			2200	

After Blair-West et al. (1968).

Table 8.2. *Physiological parameters in Eastern Grey Kangaroos and Common Wombats from a sodium-deficient alpine environment (the Snowy Mountains) and a coastal environment in Victoria*

	Snowy Mountains		Coastal Victoria	
	M. giganteus	*V. ursinus*	*M. giganteus*	*V. ursinus*
Urine sodium (mEq·l^{-1})	0	0	215	23
Plasma sodium (mEq·l^{-1})	139	NA	148	NA
Weight of adrenal glands (g)	1.1	0.7	0.6	0.4
Blood aldosterone (μg·dl^{-1})	40	12	9	6
Area of zona glomerulosa of left adrenal (mm^2)	15	14	6	6

After Blair-West et al. (1968).
NA – not available.

Wales. The Red Kangaroo was higher, reflecting the greater intake of halophytes by this species than by the Euro (Dawson & Denny, 1969). In Blair-West et al.'s (1968) study, the low value for urinary sodium in coastal *V. ursinus* was also thought to reflect differences in diet; grass roots, which are relatively low in sodium, rather than tops appeared to be eaten by *V. ursinus*.

Only limited observations on blood electrolytes were made by Blair-West et al. (1968), but the sodium content of plasma of alpine *M. giganteus* was reduced relative to the coastal animals. The adrenal glands of both *M. giganteus* and *V. ursinus* examined in the Snowy Mountains were approximately double the weight of those in coastal animals, and the area of zona glomerulosa, the site of synthesis and secretion of aldosterone, was greater in adrenal glands of both species from the sodium-deficient Snowy Mountains. Correlated with this was a higher concentration of circulatory aldosterone in both species in the alpine environment. There was little difference in the levels of the glucocorticoids cortisol and corticosterone in sodium-deficient and sodium-replete animals. Thus the several differences described are clearly adaptations specifically to achieve sodium homeostasis in a sodium-deficient environment.

Blair-West et al. (1968) also examined the salivary glands of *M. giganteus* from the two environments because of the high salivary flow rates in forestomach fermenters such as ruminants and macropodine marsupials. Maintenance of microbial fermentation in the forestomach involves circulation of large volumes of fluid of high sodium content. The capacity to replace sodium with potassium in this cycle in sodium deficiency reduces the haemodynamic stress of sodium deficiency and thus is pivotal to adaptation. There were marked structural differences in the salivary glands between the two populations of *M. giganteus*. Animals from the Snowy Mountains had a much more extensive duct system of both the parotid and submandibular glands, and blood vessels were extraordinarily abundant around the striated (secretory) ducts. These structural differences are adaptations for conserving sodium by the salivary glands. Although salivary concentrations of sodium and potassium were not measured by Blair-West et al. (1968), there would undoubtedly be a much lower sodium/potassium ratio in the saliva of alpine *M. giganteus* than in that of coastal animals.

Salivary glands from *V. ursinus*, a hindgut fermenter (Chapter 4), showed similar structural differences between the two environments though not to the same degree.

Ascorbic acid biosynthesis in marsupials and monotremes and the phylogeny of marsupials

The pathway by which ascorbic acid (vitamin C) is synthesised in those vertebrates capable of synthesising it has been extrapolated from work done with rat liver preparations (Fig. 8.1). All species known to require dietary ascorbate lack the enzyme L-gulonolactone oxidase (Chatterjee, 1973), which catalyses the final step in the biosynthetic pathway, the oxidation of L-gulonolactone to L-ascorbic acid.

In amphibians, reptiles and some birds, L-gulonolactone oxidase is located in the kidney. All species of placental mammals reported to be capable of synthesising ascorbate have L-gulonolactone oxidase solely in

Fig. 8.1. Biosynthetic pathway of ascorbic acid (vitamin C) in rat liver.

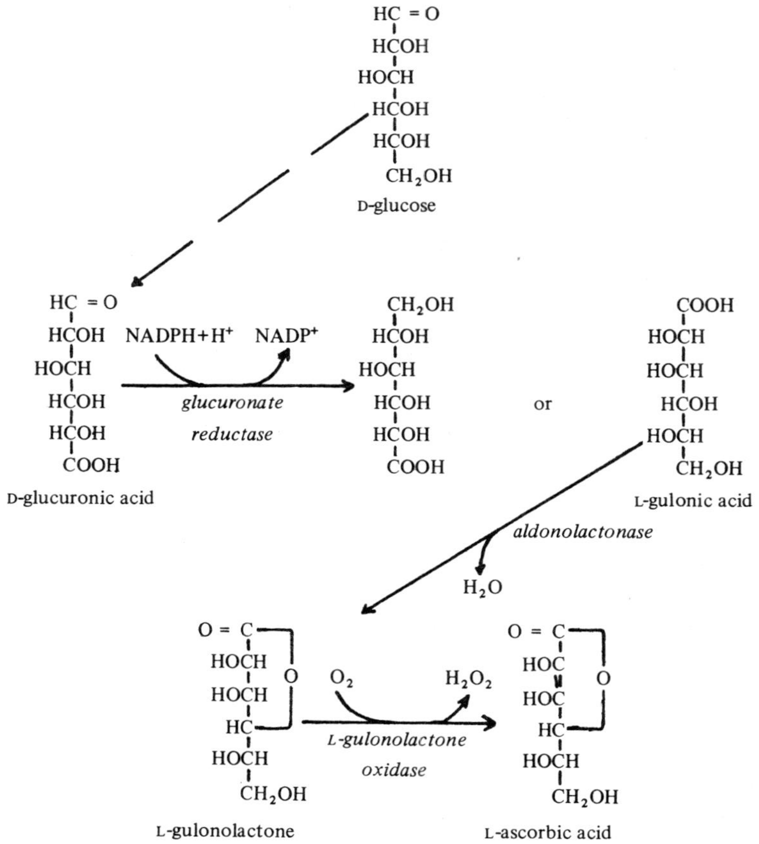

$$\begin{array}{c} HC=O \\ | \\ HCOH \\ | \\ HOCH \\ | \\ HCOH \\ | \\ HCOH \\ | \\ CH_2OH \end{array}$$
D-glucose

D-glucuronic acid:
$$\begin{array}{c} HC=O \\ | \\ HCOH \\ | \\ HOCH \\ | \\ HCOH \\ | \\ HCOH \\ | \\ COOH \end{array}$$

$\xrightarrow{\text{glucuronate reductase}}$ NADPH+H⁺ → NADP⁺

$$\begin{array}{c} CH_2OH \\ | \\ HCOH \\ | \\ HOCH \\ | \\ HCOH \\ | \\ HCOH \\ | \\ COOH \end{array}$$

or L-gulonic acid:
$$\begin{array}{c} COOH \\ | \\ HOCH \\ | \\ HOCH \\ | \\ HCOH \\ | \\ HOCH \\ | \\ CH_2OH \end{array}$$

aldonolactonase, H₂O

L-gulonolactone:
$$\begin{array}{c} O=C\!\!-\!\! \\ | \quad\quad\;| \\ HOCH \;\;O \\ | \\ HOCH \\ | \\ HC\!\!-\!\!\rfloor \\ | \\ HOCH \\ | \\ CH_2OH \end{array}$$

$\xrightarrow[\text{oxidase}]{L\text{-gulonolactone}}$ O₂ → H₂O₂

L-ascorbic acid:
$$\begin{array}{c} O=C\!\!-\!\! \\ | \quad\quad\;| \\ HOC \;\;O \\ \|\\ HOC \\ | \\ HC\!\!-\!\!\rfloor \\ | \\ HOCH \\ | \\ CH_2OH \end{array}$$

the liver (Chatterjee, 1973). The only species of marsupial investigated before 1979 was the Virginia Opossum (*Didelphis virginiana*); in this species the enzyme was reported in the liver (Nakajima, Shantha & Bourne, 1969).

Birney, Jenness & Hume (1979, 1980) have further investigated the locus of L-gulonolactone oxidase in marsupials, and also in monotremes because of their numerous reptilian characteristics (Griffiths, 1978). Birney *et al.*'s (1979, 1980) results are summarised in Table 8.3. Although the quantitative relationship between the activity of L-gulonolactone oxidase *in vitro* and

Table 8.3 *Ascorbic acid synthesis (as measured by L-gulonolactone oxidase activity in $\mu mol \cdot g^{-1}$ tissue homogenate $\cdot h^{-1}$) in the kidney and liver of monotremes and marsupials; values are means (with ranges)*

Species (with number of animals)	Kidney	Liver
Monotremata		
Ornithorhynchus anatinus (3)	10.1 (6.6–12.6)	Nil
Tachyglossus aculeatus (3)	18.8 (15.9–23.4)	Nil
Didelphidae		
Didelphis virginiana (13)	Nil	3.3 (0.4–6.9)
Dasyuridae		
Dasyuroides byrnei (2)	Nil	0.5 (0.4–0.6)
Antechinus stuartii (11)	Nil	1.2 (0.3–1.7)
Dasyurus maculatus (1)	Nil	0.9
Peramelidae		
Perameles nasuta (1)	2.8	1.7
Isoodon macrourus (3)	5.3 (4.9–6.2)	4.4 (4.3–4.4)
Petauridae		
Pseudocheirus peregrinus (7)	Nil	1.6 (0.0–3.6)
Petauroides volans (7)	Nil	0.8 (0.5–1.0)
Burramyidae		
Cercartetus nanus (1)	Nil	2.5
Vombatidae		
Vombatus ursinus (2)	Nil	5.6 (3.7–7.4)
Phalangeridae		
Trichosurus vulpecula (9)	Nil	2.5 (0.5–4.4)
Macropodidae		
Macropus rufogriseus (6)	0.6	5.2 (3.4–7.5)
Macropus eugenii (1)	Nil	1.9
Macropus giganteus (5)	Nil	2.8 (1.7–4.4)
Macropus robustus robustus (1)	Nil	1.7
Thylogale thetis (4)	0.4 (0.3–0.4)	4.2 (1.0–5.6)
Wallabia bicolor (5)	Nil	4.0 (3.6–4.5)

After Birney, Jenness & Hume (1980).

the rate of synthesis of ascorbate *in vivo* is unknown, it seems safe to assume that the enzyme activities measured reflect relative rates of ascorbate biosynthesis.

Both the Platypus (*Ornithorhynchus anatinus*) and the Echidna (*Tachyglossus aculeatus*) were found to contain L-gulonolactone oxidase only in the kidney, in the manner of reptiles. Within the kidney, the activity was largely or entirely in the cortex (Birney *et al.*, 1979). This is the first time that ascorbic acid biosynthesis has been demonstrated in a mammalian kidney.

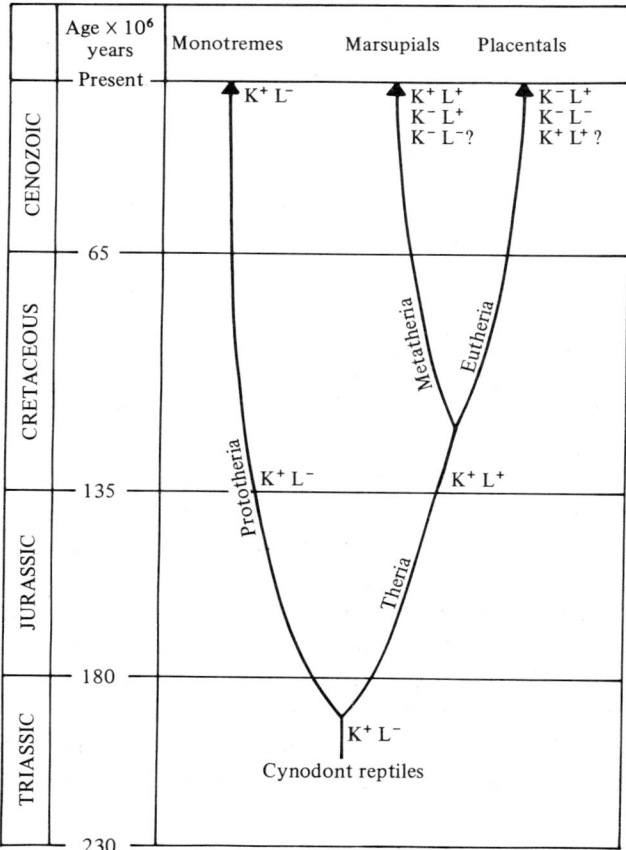

Fig. 8.2. Phylogeny of the subclasses and infraclasses of living mammals, showing the tissue locus of the enzyme L-gulonolactone oxidase in living groups and its presumed presence or absence in ancestral groups. Kidney is symbolised by K, liver by L, the presence of the enzyme by +, and its absence by −. After Birney *et al.* (1980).

Bandicoots of both genera studied, *Perameles* and *Isoodon*, had similar levels of activity in liver and kidney. Some individuals of two macropodine species, *Macropus rufogriseus* and *Thylogale thetis*, also exhibited activity in the kidney, but with relatively much higher levels in the liver. All the other marsupials studied had activity exclusively in the liver, in the manner of those eutherians capable of synthesising ascorbic acid.

From this information on monotremes and marsupials, together with what is known about ascorbic acid biosynthesis in eutherians and the fossil record, Birney *et al.* (1980) proposed the following evolutionary pathway of ascorbate synthesis in the living mammals (Fig. 8.2).

Cynodont reptiles and early mammals are hypothesised to have synthesised ascorbate in the kidney in the manner of living reptiles and monotremes, but mutation(s) during the evolution of early therians allowed them to synthesise ascorbate in both kidney and liver, as we have seen in the living bandicoots. When tissue alternatives have existed, selection apparently has generally favoured synthesis in liver alone over kidney alone or both kidney and liver. Birney *et al.* (1980) suggested several possible advantages of liver as the sole locus of ascorbate synthesis, but greater organ size and biosynthetic capacity are probably the primary advantages of liver over kidney. Cell specialisation may be the primary advantage of liver alone over liver and kidney together. The alternative hypothesis that kidney synthesis in living therians is a re-acquired characteristic was also considered by Birney *et al.* (1980), and while neither hypothesis can be rejected outright with the available data, the retention hypothesis is favoured by these workers.

The phylogeny of the living marsupials, based on Kirsch (1977a, c) is shown in the final figure (Fig. 8.3), but modified in the light of the results obtained by Birney *et al.* (1979) on ascorbate biosynthetic ability and tissue locus in the bandicoots. In fact there is only one point of contention, and that is the purported relationship of the Perameloidea (families Thylacomyidae and Peramelidae) to the Dasyuroidea. Apparently, largely on the basis of serological studies, Kirsch (1977a, c) considered the bandicoots to have shared a common ancestor with the dasyuroids into the mid-Palaeocene, well after the diprotodont lineage had taken its origin. However, Birney *et al.*'s (1980) interpretation of the comparisons of fossil marsupials by Ride (1964), of comparisons of dentition and cranial structure of Archer (1976a, b), of considerations of the peramelid tarsus by Marshall (1972), and of their own findings on L-gulonolactone oxidase activity and locus, suggests that the perameloids separated from the dasyuroids earlier than illustrated by Kirsch (Fig. 8.3). This persuaded Birney *et al.* (1980) to agree with Ride (1964) that the bandicoots differ sufficiently from the didelphoids and dasyuroids that they should be regarded as members of a distinct order, the Peramelina.

Fig. 8.3. Phylogeny of living marsupials (modified from Kirsch, 1977a, c), showing the tissue locus of the enzyme L-gulonolactone oxidase in the families examined by Birney *et al.* (1980). The presumed branching points for families not studied are shown by the dotted lines. In the Peramelidae the broken line illustrates Kirsch's interpretation and the solid line the conclusion reached by Birney *et al.* (1980). Kidney is symbolised by K, liver by L, the presence of L-gulonolactone by + and its absence by −.

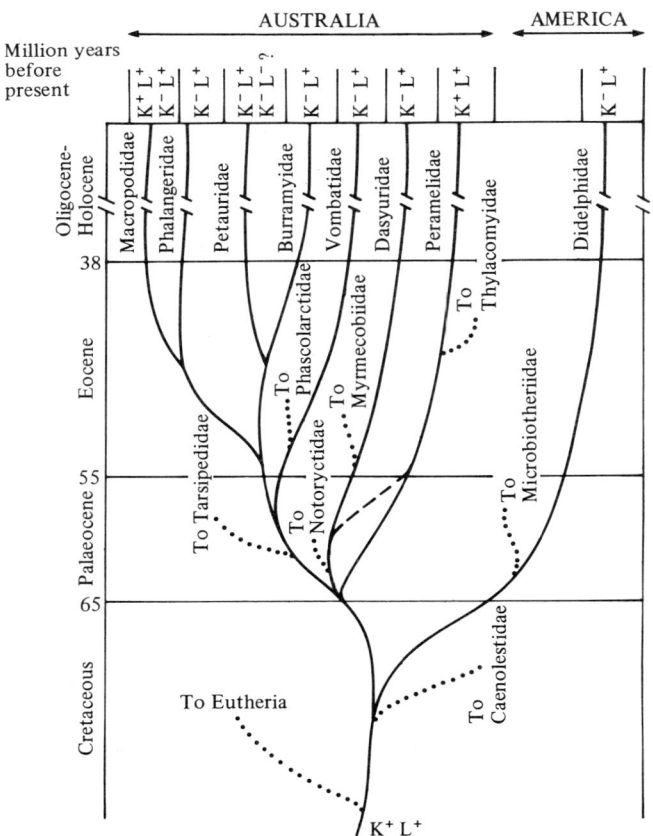

Summary

In line with most sections of this book, this final chapter has been more concerned with herbivores than with carnivores or omnivores. This is because herbivores, particularly grazers, can be subjected to much wider fluctuations in their supply of minerals and some vitamins than are the other groups.

The most definitive work on trace element nutrition of marsupials has been carried out with the Quokka (*Setonix brachyurus*). This is because of the generally poor nutritional status of the limestone-derived sandy soils of Rottnest, and of the failure of early attempts to graze domestic sheep on the Island. Sodium and potassium, two of the macro-elements, have been studied in a wide range of localities because of the problems of both sodium excess (on some island habitats) and sodium deficiency (in the Snowy Mountains).

The only vitamins to have received any attention are vitamin B_{12} (in relation to cobalt deficiency on Rottnest Island), vitamin E (in relation to nutritional muscular dystrophy in Quokkas), and vitamin C (ascorbic acid). The ability of both marsupials and monotremes to synthesise ascorbic acid, and differences in the locus of synthesis among groups of marsupials are described in relation to current hypotheses on the evolution and phylogeny of the Marsupialia.

Appendix

A classification of extant monotreme and marsupial species mentioned in the text. It is based on the classification proposed by Kirsch & Calaby (1977), synonyms in Kirsh (1977b), and common names of Australian species as recommended by the Vernacular Names Committee of the Australian Mammal Society (Strahan, 1980).

Class	Mammalia
Subclass	Prototheria
Order	Monotremata
Family	Tachyglossidae
	Tachyglossus aculeatus, Echidna
Family	Ornithorhynchidae
	Ornithorhynchus anatinus, Platypus
Subclass	Metatheria (Marsupialia)
Order	Polyprotodontia
Family	Didelphidae
	Marmosa robinsoni, Murine Opossum
	Metachirus nudicaudatus, Brown Four-eyed Opossum
	Didelphis marsupialis, Common Opossum
	Didelphis virginiana, Virginia Opossum
	Philander opossum (= *Metachirops opossum*), Grey Four-eyed Opossum
	Chironectes minimus, Water Opossum
Family	Dasyuridae
	Planigale gilesi, Paucident Planigale
	Planigale ingrami, Long-tailed Planigale
	Planigale maculatus (= *P. maculata*), Common Planigale
	Planigale tenuirostris, Narrow-nosed Planigale
	Antechinus bellus, Fawn Antechinus
	Antechinus bilarni, Sandstone Antechinus
	Antechinus flavipes, Yellow-footed Antechinus

	Antechinus macdonnellensis (= *Pseudantechinus macdonnellensis*), Fat-tailed Antechinus
	Antechinus minimus, Swamp Antechinus
	Antechinus stuartii, Brown Antechinus
	Antechinus swainsonii, Dusky Antechinus
	Phascogale tapoatafa, Brush-tailed Phascogale (Tuan)
	Dasycercus cristicauda, Mulgara
	Dasyurus geoffroii, Western Quoll
	Dasyurus hallucatus, Northern Quoll
	Dasyurus maculatus, Tiger Quoll (Tiger Cat)
	Dasyurus viverrinus, Eastern Quoll
	Sarcophilis harrisii, Tasmanian Devil
	Dasyuroides byrnei, Kowari
	Sminthopsis crassicaudata, Fat-tailed Dunnart
	Sminthopsis macroura (= *S. larapinta*, *S. froggati*), Stripe-faced Dunnart
	Antechinomys laniger, Kultarr
Family	Myrmecobiidae
	Myrmecobius fasciatus, Numbat
Family	Thylacinidae
	Thylacinus cynocephalus, Thylacine (Tasmanian Tiger)
Family	Notoryctidae
	Notoryctes typhlops, Marsupial Mole
Family	Peramelidae
	Perameles bougainville, Western Barred Bandicoot
	Perameles eremiana, Desert Bandicoot
	Perameles gunnii, Eastern Barred Bandicoot
	Perameles nasuta, Long-nosed Bandicoot
	Isoodon macrourus, Northern Brown Bandicoot (Brindled Bandicoot, Short-nosed Bandicoot)
	Isoodon obesulus, Southern Brown Bandicoot
	Chaeropus ecaudatus (= *C. castanotis*), Pig-footed Bandicoot
Family	Thylacomyidae
	Macrotis lagotis, Greater Bilby (Rabbit-eared Bandicoot)
Order	Paucituberculata
Family	Caenolestidae
	Caenolestes obscurus, Rat Opossum
Order	Diprotodontia
Family	Phalangeridae
	Trichosurus caninus, Mountain Brushtail Possum (Bobuck)
	Trichosurus vulpecula, Common Brushtail Possum
	Wyulda squamicaudata, Scaly-tailed Possum
	Phalanger maculatus, Spotted Cuscus
	Phalanger orientalis, Grey Cuscus
Family	Burramyidae
	Cercartetus nanus, Eastern Pygmy-possum
	Acrobates pygmaeus, Feathertail Glider
	Burramys parvus, Mountain Pygmy-possum

Appendix

Family   Petauridae
        Gymnobelideus leadbeateri, Leadbeater's Possum
        Petaurus australis, Yellow-bellied Glider
        Petaurus breviceps, Sugar Glider
        Pseudocheirus peregrinus, Ringtail Possum
        Petauroides volans (= *Schoinobates volans*), Greater Glider
        Dactylopsila trivirgata, Striped Possum
Family   Macropodidae
Subfamily Potoroinae
        Hypsiprymnodon moschatus, Musk Rat-kangaroo
        Potorous tridactylus, Long-nosed Potoroo
        Bettongia gaimardi, Tasmanian Bettong
        Bettongia lesueur, Burrowing Bettong (Boodie)
        Bettongia penicillata, Brush-tailed Bettong (Woylie)
        Aepyprymnus rufescens, Rufous Bettong (Rufous Rat-kangaroo)
Subfamily Macropodinae
        Thylogale billardierii, Red-bellied Pademelon
        Thylogale brunii
        Thylogale stigmatica, Red-legged Pademelon
        Thylogale thetis, Red-necked Pademelon
        Petrogale inornata
        Petrogale penicillata, Brush-tailed Rock-wallaby
        Petrogale xanthopus, Yellow-footed Rock-wallaby
        Peradorcus concinna (= *Petrogale concinna*), Little Rock-wallaby (Nabarlek)
        Lagorchestes conspicillatus, Spectacled Hare-wallaby
        Lagorchestes hirsutus, Rufous Hare-wallaby (Western Hare-wallaby)
        Setonix brachyurus, Quokka
        Lagostrophus fasciatus, Banded Hare-wallaby
        Macropus agilis, Agile Wallaby (Sandy Wallaby)
        Macropus dorsalis, Black-striped Wallaby
        Macropus eugenii (= *Protemnodon eugenii*), Tammar Wallaby (Kangaroo Island Wallaby)
        Macropus fuliginosus, Western Grey Kangaroo
        Macropus giganteus, Eastern Grey Kangaroo
        Macropus irma, Western Brush Wallaby
        Macropus parma, Parma Wallaby
        Macropus parryi, Whiptail Wallaby (Pretty-face Wallaby)
        Macropus robustus robustus, Wallaroo
        Macropus robustus erubescens, Euro
        Macropus rufogriseus (= *M. bennetti*), Red-necked Wallaby (Bennett's Wallaby)
        Macropus rufus (= *Megaleia rufa*), Red Kangaroo
        Onychogalea fraenata, Bridled Nailtail Wallaby
        Wallabia bicolor (Swamp Wallaby)
        Dendrolagus bennettianus, Bennett's Tree-kangaroo
        Dendrolagus goodfellowi, Goodfellow's Tree-kangaroo
        Dendrolagus matschiei

Dendrolagus ursinus
Dorcopsis luctuosa (= *D. muelleri luctuosa*)
Dorcopsoides
Dorcopsulus
Family Phascolarctidae
Phascolarctos cinereus, Koala
Family Vombatidae
Vombatus ursinus, Common Wombat
Lasiorhinus latifrons, Hairy-nosed Wombat
Family Tarsipedidae
Tarsipes spencerae, Honey Possum

References

Adrian, J. (1976). Gums and hydrocolloids in nutrition. *World Review of Nutrition and Dietetics*, **25**, 189–216.

Aleksiuk, M. & Cowan, I. McT. (1969). The winter metabolic depression in Arctic beavers (*Castor canadensis* Kuhl) with comparisons to California beavers. *Canadian Journal of Zoology*, **47**, 965–79.

Andrewartha, H. G. & Barker, S. (1969). Introduction to a study of the ecology of the Kangaroo Island wallaby, *Protemnodon eugenii* (Desmarest), within Flinders Chase, Kangaroo Island, South Australia. *Transactions of the Royal Society of South Australia*, **93**, 127–32.

Annison, E. F. & Armstrong, D. G. (1970). Volatile fatty acid metabolism. In *Physiology of Digestion and Metabolism in the Ruminant*, ed. A. T. Phillipson, pp. 422–37. Newcastle-upon-Tyne: Oriel Press.

Archer, M. (1976a). The dasyurid dentition and its relationships to that of didelphids, thylacinids, borhyaenids (Marsupicarnivora) and peramelids (Peramelina: Marsupialia). *Australian Journal of Zoology, Supplementary Series*, **39**, 1–34.

Archer, M. (1976b). The basicranial region of marsupicarnivores (Marsupialia), inter-relationships of carnivorous marsupials, and affinities of the insectivorous marsupial peramelids. *Journal of the Linnean Society, London (Zoology)*, **59**, 217–322.

Archer, M. (1979). The status of Australian dasyurids, thylacinids and myrmecobiids. In *The Status of Endangered Australasian Wildlife*, ed. M. J. Tyler, pp. 29–43. Adelaide: Royal Zoological Society of South Australia.

Arnold, J. & Shield, J. (1970). Oxygen consumption and body temperature of the churditch (*Dasyurus geoffroii*). *Journal of Zoology, London*, **160**, 391–404.

Bailey, P. T. (1971). The red kangaroo, *Megaleia rufa* (Desmarest), in north-western New South Wales. I. Movements. *CSIRO Wildlife Research*, **16**, 11–28.

Bailey, P. T., Martensz, P. N. & Barker, R. (1971). The red kangaroo, *Megaleia rufa* (Desmarest), in north-western New South Wales. II. Food. *CSIRO Wildlife Research*, **16**, 29–39.

Baker, R. T. & Smith, H. G. (1920). *The Eucalypts and their Essential Oils*, 2nd ed. Sydney: Government Printer.

Ballard, F. J. (1965). Glucose utilization in mammalian liver. *Comparative Biochemistry and Physiology*, **14**, 437–43.

Ballard, F. J., Hanson, R. W. & Kronfeld, D. S. (1969). Gluconeogenesis and lipogenesis in tissue from ruminant and nonruminant animals. *Federation Proceedings*, **28**, 218–31.

Barker, I. K., Beveridge, I., Bradley, A. J. & Lee, A. K. (1978). Observations on spontaneous stress-related mortality among males of the dasyurid marsupial *Antechinus stuartii* Macleay. *Australian Journal of Zoology*, **26**, 435–47.

Barker, J. M. (1961). The metabolism of carbohydrate and volatile fatty acids in the marsupial, *Setonix brachyurus*. *Quarterly Journal of Experimental Physiology*, **46**, 54–68.

Barker, S. (1960). The role of trace elements in the biology of the Quokka (*Setonix brachyurus*, Quoy & Gaimard). Ph.D. thesis, University of Western Australia, Perth.

Barker, S. (1961a). Studies on marsupial nutrition. III. The copper-molybdenum-inorganic sulphate interaction in the Rottnest quokka, *Setonix brachyurus* (Quoy & Gaimard). *Australian Journal of Biological Sciences*, **14**, 646–58.

Barker, S. (1961b). Copper, molybdenum and inorganic sulphate levels in Rottnest plants. *Journal of the Royal Society of Western Australia*, **44**, 49–52.

Barker, S. (1962). Copper-levels in the milk of a marsupial. *Nature*, **193**, 292.

Barker, S. (1968). Nitrogen balance and water intake in the Kangaroo Island wallaby, *Protemnodon eugenii* (Desmarest). *Australian Journal of Experimental Biology and Medical Science*, **46**, 17–32.

Barker, S. (1971). Nitrogen and water excretion of wallabies: differences between field and laboratory findings. *Comparative Biochemistry and Physiology*, **38A**, 359–67.

Barker, S. (1974). Studies on seasonal anaemia in the Rottnest Island quokka, *Setonix brachyurus* (Quoy & Gaimard) (Marsupialia: Macropodidae). *Transactions of the Royal Society of South Australia*, **98**, 43–8.

Barker, S., Brown, G. D. & Calaby, J. H. (1963). Food regurgitation in the Macropodidae. *Australian Journal of Science*, **25**, 430–2.

Barker, S., Glover, R., Jacobsen, P. & Kakulas, B. A. (1974). Seasonal anaemia in the Rottnest quokka, *Setonix brachyurus* (Quoy & Gaimard) (Marsupialia: Macropodidae). *Comparative Biochemistry and Physiology*, **49A**, 147–58.

Barker, S., Lintern, S. M. & Murphy, C. R. (1970). The effect of water restriction on urea retention and nitrogen excretion in the Kangaroo Island wallaby, *Protemnodon eugenii* (Desmarest). *Comparative Biochemistry and Physiology*, **34**, 883–93.

Barnard, E. A. (1969). Biological function of pancreatic ribonuclease. *Nature*, **221**, 340–4.

Barnes, D. B. (1977). The special anatomy of *Marmosa robinsoni*. In *The Biology of Marsupials*, ed. D. Hunsaker, II, pp. 387–413. New York: Academic Press.

Barnett, J. L., How, R. A. & Humphreys, W. F. (1979a). Blood parameters in natural populations of *Trichosurus* species (Marsupialia: Phalangeridae). I. Age, sex and seasonal variation in *T. caninus* and *T. vulpecula*. *Australian Journal of Zoology*, **27**, 913–26.

Barnett, J. L., How, R. A. & Humphreys, W. F. (1979b). Blood parameters in natural populations of *Trichosurus* species (Marsupialia: Phalangeridae). II. Influence of habitat and population strategies of *T. caninus* and *T. vulpecula*. *Australian Journal of Zoology*, **27**, 927–38.

Barnett, J. L. (1974). Changes in the hydroxyproline concentration of the skin of *Antechinus stuartii* with age and hormonal treatment. *Australian Journal of Zoology*, **22**, 311–18.

Bartholomai, A. (1972). Aspects of the evolution of the Australian marsupials. *Proceedings of the Royal Society of Queensland*, **82**, 5–18.

Bauchop, T. (1979a). The rumen anaerobic fungi: colonizers of plant fibre. *Annales de Recherches Veterinaires*, **10**, 246–8.

Bauchop, T. (1979b). Rumen anaerobic fungi of cattle and sheep. *Applied and Environmental Microbiology*, **37**, 148–58.

Baudinette, R. V., Wheldrake, J. F., Hewitt, S. & Hawke, D. (1980). The metabolism of [^{14}C]phenol by native Australian rodents and marsupials. *Australian Journal of Zoology*, **28**, 511–20.

Bauman, T. R. & Turner, C. W. (1966). L-Thyroxine secretion rates and L-triiodithyronine equivalents in the opossum (*Didelphis virginianus*). *General and Comparative Endocrinology*, **6**, 109–13.

Bell, H. M. (1973). The ecology of three macropod marsupial species in an area of open forest and savannah woodland in north Queensland, Australia. *Mammalia*, **37**, 527–44.

Bell, R. H. V. (1969). The use of the herb layer by grazing ungulates in the Serengeti. In *Animal Populations in Relation to their Food Resources*, ed. A. Watson, pp, 111–28. Oxford: Blackwell.

Bensley, R. R. (1902). The cardiac glands of mammals. *American Journal of Anatomy*, **2**, 105–65.

Betts, T. J. (1978). Koala acceptance of *Eucalyptus globulus* labill as food in relation to the proportion of sesquiterpenoids in the leaves. In *The Koala*, ed. T. J. Bergin, pp. 75–85. Sydney: Zoological Parks Board of New South Wales.

Birney, E. C., Jenness, R. & Hume, I. D. (1979). Ascorbic acid biosynthesis in the mammalian kidney. *Experientia*, **35**, 1425.

Birney, E. C., Jenness, R. & Hume, I. D. (1980). Evolution of an enzyme system: Ascorbic acid biosynthesis in monotremes and marsupials. *Evolution*, **34**, 230–9.

Björnhag, G. (1972). Separation and delay of contents in the rabbit colon. *Swedish Journal of Agricultural Research*, **2**, 125–36.

Björnhag, B. & Sperber, I. (1977). Transport of various food components through the digestive tract of turkeys, geese and Guinea fowl. *Swedish Journal of Agricultural Research*, **7**, 57–66.

Black, J. L. (1971). A theoretical consideration of the effect of preventing rumen fermentation on the efficiency of utilization of dietary energy and protein in lambs. *British Journal of Nutrition*, **25**, 31–55.

Blackhall, S. (1980). Diet of the eastern native-cat, *Dasyurus viverrinus* (Shaw), in southern Tasmania. *Australian Wildlife Research*, **7**, 191–7.

Blair-West, J. R., Coghlan, J. P., Denton, D. A., Nelson, J. F., Orchard, E., Scoggins, B. A., Wright, R. D., Myers, K. & Junqueira, C. L. (1968). Physiological, morphological and behavioural adaptation to a sodium deficient environment by wild native Australian and introduced species of animals. *Nature*, **217**, 922–8.

Blaxter, K. L., Graham, N. McC. & Wainman, F. W. (1956). Some observations on the digestibility of food by sheep, and on related problems. *British Journal of Nutrition*, **10**, 69–91.

Bligh, J. & Johnson, K. G. (1973). Glossary of terms for thermal physiology. *Journal of Applied Physiology*, **35**, 941–61.

Bolton, B. L. & Latz, P. K. (1978). The western hare-wallaby, *Lagorchestes hirsutus* (Gould) (Macropodidae), in the Tanami Desert. *Australian Wildlife Research*, **5**, 285–93.

Bradley, A. J., McDonald, I. R. & Lee, A. K. (1975). Effects of exogenous cortisol on mortality of a dasyurid marsupial. *Journal of Endocrinology*, **66**, 281–2.

Bradley, A. J., McDonald, I. R. & Lee, A. K. (1976). Corticosteroid binding globulin and mortality in a dasyurid marsupial. *Journal of Endocrinology*, **70**, 323–4.

Brody, S. (1945). *Bioenergetics and Growth*. New York: Reinhold.

Brody, S. & Procter, R. C. (1932). Relation between basal metabolism and mature body weight in different species of mammals and birds. *Missouri Agricultural Experiment Station Research Bulletin*, **166**, 89–101.

Brown, G. D. (1964). The nitrogen requirements of macropod marsupials. Ph.D. thesis, University of Western Australia, Perth.

Brown, G. D. (1968). The nitrogen and energy requirements of the euro (*Macropus robustus*) and other species of macropod marsupials. *Proceedings of the Ecological Society of Australia*, **3**, 106–12.

Brown, G. D. (1969). Studies on marsupial nutrition. VI. The utilization of dietary urea by the euro or kill kangaroo, *Macropus robustus* (Gould). *Australian Journal of Zoology*, **17**, 187–94.

Brown, G. D. & Main, A. R. (1967). Studies on marsupial nutrition. V. The nitrogen requirements of the euro, *Macropus robustus*. *Australian Journal of Zoology*, **15**, 7–27.

Brunner, H. & Cowan, B. J. (1974). *The Identification of Mammalian Hair*. Melbourne: Inkata Press.

Buchmann, O. L. K. & Guiler, E. R. (1977). Behaviour and ecology of the Tasmanian devil, *Sarcophilus harrisii*. In *The Biology of Marsupials*, ed. B. Stonehouse & D. Gilmore, pp. 155–68. London: Macmillan.

Calaby, J. H. (1958). Studies in marsupial nutrition. II. The rate of passage of food residues and digestibility of crude fibre and protein by the quokka, *Setonix brachyurus* (Quoy and Gaimard). *Australian Journal of Biological Sciences*, **11**, 571–80.

Calaby, J. H. (1960). Observations on the banded ant-eater *Myrmecobius f. fasciatus*. Waterhouse (Marsupialia), with particular reference to its food habits. *Proceedings of the Zoological Society of London*, **135**, 183–207.

Calaby, J. H. (1966). Mammals of the upper Richmond and Clarence Rivers, New South Wales. *Technical Paper of the Division of Wildlife Survey, CSIRO Australia*, **10**, 1–55.

Carlsson, A. (1915). Zur Morphologie des *Hypsiprymnodon moschatus*. *Kungliga Svenska Vetenskapsakademiens Handlingar*, **52**, 1–48.

Carroll, E. J. & Hungate, R. E. (1954). The magnitude of the microbial fermentation in the bovine rumen. *Applied Microbiology*, **2**, 205–14.

Casey, T. M. & Casey, K. K. (1979). Thermoregulation of Arctic weasels. *Physiological Zoology*, **52**, 153–64.

Castle, E. J. (1956). The rate of passage of foodstuffs through the alimentary tract of the goat. I. Studies on adult animals fed on hay and concentrates. *British Journal of Nutrition*, **10**, 15–23.

Catling, P. C. & Vinson, G. P. (1976). Adrenocortical hormones in the neonate and pouch young of the tammar wallaby, *Macropus eugenii*. *Journal of Endocrinology*, **69**, 447–8.

Caughley, G. J. (1964). Density and dispersion of two species of kangaroo in relation to habitat. *Australian Journal of Zoology*, **12**, 238–49.

Charles-Dominique, P. (1974). Ecology and feeding behaviour of sympatric lorisids in Gabon. In *Prosimian Biology*, ed. R. D. Martin, G. A. Doyle & A. C. Walker. Pittsburgh: University of Pittsburgh Press.

Chatterjee, I. B. (1973). Evolution and the biosynthesis of ascorbic acid. *Science*, **182**, 1271–2.

Cheal, P. D., Lee, A. K. & Barnett, J. L. (1976). Changes in the haematology of *Antechinus stuartii* (Marsupialia), and their association with male mortality. *Australian Journal of Zoology*, **24**, 299–311.

Christensen, P. E. S. (1977). The biology of *Bettongia penicillata* and *Macropus eugenii* in relation to fire. Ph.D. thesis, University of Western Australia, Perth.

Clemens, E. T. & Stevens, C. E. (1979). Sites of organic acid production and patterns of digesta movement in the gastrointestinal tract of the raccoon. *Journal of Nutrition*, **109**, 1110–16.

Cocimano, M. R. & Leng, R. A. (1967). Metabolism of urea in sheep. *British Journal of Nutrition*, **21**, 353–71.

Coghlan, J. P. & Scoggins, B. A. (1967). The measurement of aldosterone, cortisol and corticosterone in the blood of the wombat (*Vombatus hirsutus* Perry) and the kangaroo (*Macropus giganteus*). *Journal of Endocrinology*, **39**, 445–8.

Cook, L. J., Scott, T. W., Ferguson, K. A. & McDonald, I. W. (1970). Production of poly-unsaturated ruminant body fats. *Nature*, **228**, 178.

Cooley, H. & Janssens, P. A. (1977). Metabolic effects of infusion of cortisol and adrenocorticotrophin in the tammar wallaby (*Macropus eugenii* Desmarest). *General and Comparative Endocrinology*, **33**, 352–8.

Cork, S. J. (1981). Digestion and metabolism in the Koala (*Phascolarctos cinereus*), an arboreal folivore. Ph.D. thesis, University of New South Wales, Sydney.

Cork, S. J. & Hume, I. D. (1978). Volatile fatty acid production rates in the caecum of the greater glider. *Bulletin of the Australian Mammal Society*, **5**, 24–5 (Abstract).

Cork, S. J. & Hume, I. D. (1980). The relative importance of fermentative and non-fermentative digestion in the energy economy of the Koala (*Phascolarctos cinereus*). *Bulletin of the Australian Mammal Society*, **6**, 28–9 (Abstract).

Cork, S. J., Warner, A. C. I. & Harrop, C. J. F. (1977). Preliminary study of the rate of passage of digesta through the gut of the koala. *Bulletin of the Australian Mammal Society*, **4**, 24 (Abstract).

Cowan, I. McT., O'Riordan, A. M. & Cowan, J. S. McT. (1974). Energy requirements of the dasyurid marsupial mouse *Antechinus swainsonii* (Waterhouse). *Canadian Journal of Zoology*, **52**, 269–75.

Crowcroft, P. & Godfrey, G. K. (1968). The daily cycle of activity in two species of *Sminthopsis* (Marsupialia: Dasyuridae). *Journal of Animal Ecology*, **37**, 63–73.

Czerkawski, J. W. (1978). Reassessment of efficiency of synthesis of microbial matter in the rumen. *Journal of Dairy Science*, **61**, 1261–73.

Darlington, J. M. & Hershberger, T. V. (1968). Effect of forage maturity on digestibility, intake and nutritive value of alfalfa, timothy and orchardgrass by equine. *Journal of Animal Science* **27**, 1572–6.

Dawson, T. J. & Brown, G. D. (1970). A comparison of the insulative and reflective properties of the fur of desert kangaroos. *Comparative Biochemistry and Physiology*, **37**, 23–38.

Dawson, T. J. & Degabriele, R. (1973). The cuscus (*Phalanger maculatus*) – a marsupial sloth? *Journal of Comparative Physiology*, **83**, 41–50.

Dawson, T. J. & Denny, M. J. S. (1969). Seasonal variation in the plasma and urine electrolyte concentration of the arid zone kangaroos *Megaleia rufa* and *Macropus robustus*. *Australian Journal of Zoology*, **17**, 777–84.

Dawson, T. J., Denny, M. J. S. & Hulbert, A. J. (1969). Thermal balance of the macropodid marsupial *Macropus eugenii* Desmarest. *Comparative Biochemistry and Physiology*, **31**, 645–53.

Dawson, T. J. & Ellis, B. A. (1979). Comparison of the diets of yellow-footed rock wallabies and sympatric herbivores in western New South Wales. *Australian Wildlife Research*, **6**, 245–54.

Dawson, T. J. & Hulbert, A. J. (1969). Standard energy metabolism of marsupials. *Nature*, **222**, 383.

Dawson, T. J. & Hulbert, A. J. (1970). Standard metabolism, body temperature, and surface areas of Australian marsupials. *American Journal of Physiology*, **218**, 1233–8.

Dawson, T. J. & Wolfers, J. M. (1978). Metabolism, thermoregulation and torpor in shrew sized marsupials of the genus *Planigale*. *Comparative Biochemistry and Physiology*, **59A**, 305–9.

Dawson, W. R. (1955). The relation of oxygen consumption to temperature in desert rodents. *Journal of Mammalogy*, **36**, 543–53.

Dawson, W. R. & Bennett, A. F. (1978). Energy metabolism and thermoregulation of the spectacled hare wallaby (*Lagorchestes conspicillatus*). *Physiological Zoology*, **51**, 114–30.

Degabriele, R. (1977). The environmental biology of the koala. Master of Science Thesis, University of New South Wales, Sydney.

Degabriele, R. & Dawson, T. J. (1979). Metabolism and heat balance in an arboreal marsupial, the koala (*Phascolarctos cinereus*). *Journal of Comparative Physiology*, **134**, 293–301.

Degabriele, R., Harrop, C. J. F. & Dawson, T. J. (1978). Water metabolism of the koala

(*Phascolarctos cinereus*). In *The Ecology of Arboreal Folivores*, ed. G. G. Montgomery, pp. 163–72. Washington, DC: Smithsonian Institution Press.

Dellow, D. W. (1979). Physiology of digestion in the macropodine marsupials. Ph.D. thesis, University of New England, Armidale.

Denny, M. J. S. & Dawson, T. J. (1973). A field technique for studying water metabolism of large marsupials. *Journal of Wildlife Management*, **37**, 574–8.

Denny, M. J. S. & Dawson, T. J. (1975a). Comparative metabolism of tritiated water by macropodid marsupials. *American Journal of Physiology*, **228**, 1794–9.

Denny, M. J. S. & Dawson, T. J. (1975b). Effects of dehydration on body-water distribution in desert kangaroos. *American Journal of Physiology*, **229**, 251–4.

Denny, M. J. S. & Dawson, T. J. (1977). Kidney structure and function of desert kangaroos. *Journal of Applied Physiology*, **42**, 636–42.

Dick, A. T., Dewey, D. W. & Gawthorne, J. M. (1975). Thiomolybdates and the copper-molybdenum-suphur interaction in ruminant nutrition. *Journal of Agricultural Science*, **85**, 567–8.

Downes, A. M. & McDonald, I. W. (1964). The chromium-51 complex of ethylene diamine tetraacetic acid as a soluble rumen marker. *British Journal of Nutrition*, **18**, 153–62.

Duncan, P. E. (1979). The biology of the honey possum (*Tarsipes spencerae*) (Gray, 1842). Unpublished thesis, Murdoch University, Perth.

Du Toit, B. A. & Smuts, D. B. (1941). The endogenous nitrogen metabolism of pigs with special references to the maintenance requirement of protein. *Onderstepoort Journal of Veterinary Science*, **16**, 169–79.

Ealey, E. H. M. (1962). Biology of the euro, *Macropus robustus cervinus* (Thomas). Ph.D. thesis, University of Western Australia, Perth.

Ealey, E. H. M. (1967a). Ecology of the euro, *Macropus robustus* (Gould) in north-western Australia. I. The environment and changes in euro and sheep populations. *CSIRO Wildlife Research*, **12**, 9–25.

Ealey, E. H. M. (1967b). Ecology of the euro, *Macropus robustus* (Gould), in north-western Australia. II. Behaviour, movements and drinking patterns. *CSIRO Wildlife Research*, **12**, 27–51.

Ealey, E. H. M. & Main, A. R. (1967). Ecology of the euro, *Macropus robustus* (Gould), in north-western Australia. III. Seasonal changes in nutrition. *CSIRO Wildlife Research*, **12**, 53–65.

Eberhard, I. H., McNamara, J., Pearse, R. J. & Southwell, I. A. (1975). Ingestion and excretion of *Eucalyptus punctata* D.C. and its essential oils by the koala, *Phascolarctos cinereus* (Goldfuss). *Australian Journal of Zoology*, **23**, 169–79.

Edwards, G. P. & Ealey, E. H. M. (1975). Aspects of the ecology of the swamp wallaby *Wallabia bicolor* (Marsupialia: Macropodidae). *Australian Mammalogy*, **1**, 307–18.

Ellis, B. A., Russell, E. M., Dawson, T. J. & Harrop, C. J. F. (1977). Seasonal changes in diet preferences of free-ranging red kangaroos, euros and sheep in western New South Wales. *Australian Wildlife Research*, **4**, 127–44.

Ellis, W. C. & Huston, J. E. (1968). ^{144}Ce–^{144}Pr as a particulate digesta flow marker in ruminants. *Journal of Nutrition*, **95**, 67–78.

Enders, R. K. (1935). Mammalian life histories from Barro Colorado Island, Panama. *Bulletin of the Museum of Comparative Zoology*, **78**, 383–502.

Engelhardt, W. v., Wolter, S., Lawrenz, H. & Hemsley, J. A. (1978). Production of methane in two non-ruminant herbivores. *Comparative Biochemistry and Physiology*, **60**, 309–11.

Faichney, G. J. (1968). The production and absorption of volatile fatty acids from the rumen of the sheep. *Australian Journal of Agricultural Research*, **19**, 791–802.

Faichney, G. J. (1969). Production of volatile fatty acids in the sheep caecum. *Australian Journal of Agricultural Research*, **20**, 491–8.

Faichney, G. J. (1975). The use of markers to partition digestion within the gastro-intestinal tract of ruminants. In *Digestion and Metabolism in the Ruminant*, ed. I. W. McDonald & A. C. I. Warner, pp. 277–91. Armidale: University of New England Publishing Unit.

Farrell, D. J. & Wood, A. J. (1968). The nutrition of the female mink (*Mustela vison*). II. The energy requirements for maintenance. *Canadian Journal of Zoology*, **46**, 47–52.

Finlayson, H. H. (1958). On Central Australian mammals (with notice of related species for adjacent tracts). III. The Potoroinae. *Records of the South Australian Museum*, **13**, 235–302.

Finnemore, H., Reichard, S. K. & Large, D. K. (1935). Cyanogenetic glucosides in Australian plants. III. *Eucalyptus cladocalyx*. *Journal and Proceedings of the Royal Society of New South Wales*, **69**, 209–14.

Fitch, H. S. & Sandidge, L. L. (1953). Ecology of the opossum on a natural area in northeastern Kansas. *Publications of the Museum of Natural History of the University of Kansas*, **7**, 305–38.

Fleming, M. R. (1980). Thermoregulation and torpor in the sugar glider, *Petaurus breviceps* (Marsupialia: Petauridae). *Australian Journal of Zoology*, **28**, 521–34.

Fletcher, H. L. (1977). Habitat relationships among small mammals at Petroi, north-eastern New South Wales. Master of Natural Resources thesis, University of New England, Armidale.

Flower, W. H. (1872). Lectures on the comparative anatomy of the organs of digestion in the mammals. *Medical Times & Gazette*, **1**, 215–678.

Foley, W. J., Hume, I. D. & Taylor, R. (1980). Protein intake and requirements of the eastern wallaroo and the eastern grey kangaroo. *Bulletin of the Australian Mammal Society*, **6**, 34–35 (Abstract).

Foot, J. Z. & Romberg, B. (1965). The utilization of roughage by sheep and the red kangaroo, *Macropus rufus* (Desmarest). *Australian Journal of Agricultural Research*, **16**, 429–35.

Forbes, D. K. & Tribe, D. E. (1969). Salivary glands of kangaroos. *Australian Journal of Zoology*, **17**, 765–75.

Forbes, D. K. & Tribe, D. E. (1970). The utilization of roughages by sheep and kangaroos. *Australian Journal of Zoology*, **18**, 247–56.

Fox, L. R. & Macauley, B. J. (1977). Insect grazing on *Eucalyptus* in response to variation in leaf tannins and nitrogen. *Oecologia*, **29**, 145–62.

Fraser, E. H. & Kinnear, J. E. (1969). Urinary creatinine excretion by macropod marsupials. *Comparative Biochemistry and Physiology*, **28**, 685–92.

Freeland, W. J. & Janzen, D. H. (1974). Strategies in herbivory by mammals: The role of plant secondary compounds. *The American Naturalist*, **108**, 269–89.

Freeland, W. J. & Winter, J. W. (1975). Evolutionary consequences of eating: *Trichosurus vulpecula* (Marsupialia) and the genus *Eucalyptus*. *Journal of Chemical Ecology*, **1**, 439–55.

Frith, H. J. (1964). Mobility of the red kangaroo, *Megaleia rufa*. *CSIRO Wildlife Research*, **9**, 1–19.

Frith, H. J. & Sharman, G. B. (1964). Breeding in wild populations of the red kangaroo, *Megaleia rufa*. *CSIRO Wildlife Research*, **9**, 86–114.

Gardner, A. L. (1973). *Systematics of the Genus Didelphis*. Special Publication of the Museum of Texas Technical University, Lubbock, Texas.

Gemmell, R. T. & Engelhardt, W. v. (1977). The structure of the cells lining the stomach of the tammar wallaby (*Macropus eugenii*). *Journal of Anatomy*, **123**, 723–33.

George, G. G. (1979). The status of endangered Papua New Guinea mammals. In *The Status of Endangered Australasian Wildlife*, ed. M. J. Tyler, pp. 93–100. Adelaide: Royal Zoological Society of South Australia.

Gilmore, D. P. (1967). Foods of the Australian opossum (*Trichosurus vulpecula* Kerr) on Banks Peninsula, Canterbury, and a comparison with other selected areas. *New Zealand Journal of Science*, **10**, 235–79.

Gilmore, D. P. (1970). The rate of passage of food in the brush-tailed possum, *Trichosurus vulpecula*. *Australian Journal of Biological Sciences*, **23**, 515–18.

Godfrey, G. K. (1968). Body-temperature and torpor in *Sminthopsis crassicaudata* and *S. larapinta* (Marsupialia: Dasyuridae). *Journal of Zoology, London*, **156**, 499–511.

Gowland, P. N. (1973). Aspects of the digestive physiology of the common wombat, *Vombatus ursinus* (Shaw, 1800). Unpublished thesis, Australian National University, Canberra.

Grant, T. R. & Dawson, T. J. (1978). Temperature regulation in the platypus, *Ornithorhynchus anatinus*: Production and loss of metabolic heat in air and water. *Physiological Zoology*, **51**, 315–32.

Gray, F. V., Weller, R. A., Pilgrim, A. F. & Jones, G. B. (1967). Rates of production of volatile fatty acids in the rumen. V. Evaluation of fodders in terms of volatile fatty acid produced in the rumen of sheep. *Australian Journal of Agricultural Research*, **18**, 625–34.

Green, B. & Eberhard, I. (1979). Energy requirements and sodium and water turnovers in two captive marsupial carnivores: the Tasmanian devil, *Sarcophilus harrisii*, and the native cat, *Dasyurus viverrinus*. *Australian Journal of Zoology*, **27**, 1–8.

Green, R. H. (1967). Notes on the devil (*Sacrophilus harrisii*) and the quoll (*Dasyurus viverrinus*) in north eastern Tasmania. *Records of the Queen Victoria Museum*, **27**, 1–13.

Griffiths, M. (1978). *The Biology of the Monotremes*. New York: Academic Press.

Griffiths, M. & Barker, R. (1966). The plants eaten by sheep and by kangaroos grazing together in a paddock in south-western Queensland. *CSIRO Wildlife Research*, **11**, 145–67.

Griffiths, M., Barker, R. & MacLean, L. (1974). Further observations on the plants eaten by kangaroos and sheep grazing together in a paddock in south-western Queensland. *Australian Wildlife Research*, **1**, 27–43.

Griffiths, M. & Barton, A. A. (1966). The ontogeny of the stomach in the pouch young of the red kangaroo. *CSIRO Wildlife Research*, **11**, 169–85.

Griffiths, M., McIntosh, D. L. & Leckie, R. M. C. (1969). The effects of cortisone on nitrogen balance and glucose metabolism in diabetic and normal kangaroos, sheep and rabbits. *Journal of Endocrinology*, **44**, 1–12.

Guiler, E. R. (1970). Observations on the Tasmanian devil, *Sarcophilus harrisii* (Marsupialia: Dasyuridae). I. Numbers, home range, movements, and food in two populations. *Australian Journal of Zoology*, **18**, 49–62.

Guiler, E. R. (1971). Food of the potoroo (Marsupialia, Macropodidae). *Journal of Mammalogy*, **52**, 232–4.

Haines, H., Macfarlane, W. V., Setchell, C. & Howard, B. (1974). Water turnover and pulmocutaneous evaporation of Australian desert dasyurids and murids. *American Journal of Physiology*, **227**, 958–63.

Hall, S. (1980). The diets of two coexisting species of *Antechinus* (Marsupialia: Dasyuridae). *Australian Wildlife Research*, **7**, 365–78.

Harrington, J. (1976). The diet of the swamp wallaby, *Wallabia bicolor*, at Diamond Flat, New South Wales. Unpublished thesis, University of New England, Armidale.

Harris, L. E. & Mitchell, H. H. (1941). The value of urea in the synthesis of protein in the paunch of the ruminant. *Journal of Nutrition*, **22**, 167–82.

Harrop, C. J. F. & Barker, S. (1972). Blood chemistry and gastro-intestinal changes in the developing red kangaroo (*Megaleia rufa*, Desmarest). *Australian Journal of Experimental Biology and Medical Science*, **50**, 245–9.

Harrop, C. J. F. & Degabriele, R. (1976). Digestion and nitrogen metabolism in the koala, *Phascolarctos cinereus*. *Australian Journal of Zoology*, **24**, 201–15.

Harrop, C. J. F. & Hume, I. D. (1980). The digestive tract and digestive function in monotremes and non-macropod marsupials. In *Comparative Physiology: Primitive Mammals*, ed. K. Schmidt-Nielsen, L. Bolis & C. R. Taylor, pp. 63–77. Cambridge University Press.

Hartman, C. G. (1952). *Possums*. Austin: University of Texas Press.

Hartman, L., Shorland, F. B. & MacDonald, I. R. (1955). The trans-unsaturated acid contents of fats of ruminants and non-ruminants. *Biochemical Journal*, **61**, 603–7.

Hartnell, G. F. & Satter, L. D. (1979). Extent of particulate marker (samarium, lanthanum and cerium) movement from one digesta particle to another. *Journal of Animal Science*, **48**, 375–80.

Heighway, F. R. (1939). The anatomy of *Hypsiprymnodon moschatus*. Unpublished thesis, University of Sydney, Sydney.

Heinsohn, G. E. (1966). Ecology and reproduction of the Tasmanian bandicoots (*Perameles gunnii* and *Isoodon obesulus*). *University of California Publications in Zoology*, **80**, 1–107.

Heinsohn, G. E. (1968). Habitat requirements and reproductive potential of the macropod marsupial *Potorous tridactylus* in Tasmania. *Mammalia*, **32**, 30–43.

Hemmingsen, A. M. (1960). Energy metabolism as related to body size and respiratory surfaces, and its evolution. *Reports of the Steno Memorial Hospital and Norwegian Insulin Laboratory*, **9**, 1–110.

Henning, S. J. & Hird, F. J. R. (1970). Concentrations and metabolism of volatile fatty acids in the fermentative organs of two species of kangaroo and the guinea-pig. *British Journal of Nutrition*, **24**, 145–55.

Henning, S. J. & Hird, F. J. R. (1972). Diurnal variations in the concentrations of volatile fatty acids in the alimentary tracts of wild rabbits. *British Journal of Nutrition*, **27**, 57–64.

Herrick, E. H. (1961). Some preliminary experiments on adrenal function during seasonal stresses in a wild marsupial (*Setonix brachyurus*). *Journal of the Royal Society of Western Australia*, **44**, 61–4.

Hildwein, G. & Goffart, M. (1975). Standard metabolism and thermoregulation in a prosimian *Perodicticus potto*. *Comparative Biochemistry and Physiology*, **50A**, 201–13.

Hingson, D. J. & Milton, G. W. (1968). The mucosa of the stomach of the wombat (*Vombatus hirsutus*) with special reference to the cardiogastric gland. *Proceedings of the Linnean Society of New South Wales*, **93**, 69–75.

Hinks, N. T. & Bolliger, A. (1957). Glucuronuria in a herbivorous marsupial *Trichosurus vulpecula*. *Australian Journal of Experimental Biology and Medical Science*, **35**, 37–44.

Holleman, D. F. & Dieterich, R. A. (1973). Body water content and turnover in several species of rodents as evaluated by the tritiated water method. *Journal of Mammalogy*, **54**, 456–65.

Home, E. (1808). An account of some peculiarities in the anatomical structure of the wombat, with observations on the female organs of generation. *Philosophical Transactions of the Royal Society*, **98**, 304–12.

Home, E. (1814). *Lectures on Comparative Anatomy*, vol. I. London: I. W. Bulmer & Co.

Honigmann, H. (1936). Studies on nutrition of mammals. Part 1. *Proceedings of the Zoological Society of London, Series A*, **106**, 517–30.

Honigmann, H. (1941). Studies on nutrition of mammals. Part 3. X. Experiments with Australian silver-grey opossums (*Trichosurus vulpecula*). *Proceedings of the Zoological Society of London, Series A*, **111**, 1–35.

Houpt, T. R. (1970). Transfer of urea and ammonia to the rumen. In *Physiology of Digestion and Metabolism in the Ruminant*, ed. A. T. Phillipson, pp. 119–31. Newcastle-upon-Tyne: Oriel Press.

Hudson, J. W. (1962). The role of water in the biology of the antelope ground squirrel *Citellus leucurus*. *University of California Publications in Zoology*, **64**, 1–56.

Hudson, J. W. & Bartholomew, G. A. (1964). Terrestrial animals in dry heat: estivators. In *Handbook of Physiology*, Section 4, Chapter 34, pp. 541–50. Washington, DC: American Physiologial Society.

Hudson, J. W. & Dawson, T. J. (1975). Role of sweating from the tail in the thermal balance of the rat-kangaroo *Potorous tridactylus*. *Australian Journal of Zoology*, **23**, 453–61.

Hulbert, A. J. & Dawson, T. J. (1974a). Standard metabolism and body temperature of perameloid marsupials from different environments. *Comparative Biochemistry and Physiology*, **47A**, 583–90.

Hulbert, A. J. & Dawson, T. J. (1974b). Water metabolism in perameloid marsupials from different environments. *Comparative Biochemistry and Physiology*, **47A**, 617–33.

Hulbert, A. J. & Gordon, G. (1972). Water metabolism of the bandicoot *Isoodon macrourus* Gould, in the wild. *Comparative Biochemistry and Physiology*, **41A**, 27–34.

Hume, I. D. (1974). Nitrogen and sulphur retention and fibre digestion by euros, red kangaroos and sheep. *Australian Journal of Zoology*, **22**, 13–23.

Hume, I. D. (1977a). Production of volatile fatty acids in two species of wallaby and in sheep. *Comparative Biochemistry and Physiology*, **56A**, 299–304.

Hume, I. D. (1977b). Maintenance nitrogen requirements of the macropod marsupials *Thylogale thetis*, red-necked pademelon, and *Macropus eugenii*, tammar wallaby. *Australian Journal of Zoology*, **25**, 407–17.

Hume, I. D. (1978). Evolution of the Macropodidae digestive system. *Australian Mammalogy*, **2**, 37–42.

Hume, I. D. & Dellow, D. W. (1980a). Form and function of the macropod marsupial digestive tract. In *Comparative Physiology: Primitive Mammals*, ed. K. Schmidt-Nielsen, L. Bolis & C. R. Taylor, pp. 78–89. Cambridge University Press.

Hume, I. D. & Dellow, D. W. (1980b). Field and laboratory estimates of fermentation rates in the digestive tract of the eastern grey kangaroo. *Bulletin of the Australian Mammal Society*, **6**, 43 (Abstract).

Hume, I. D. & Dunning, A. (1979). Nitrogen and electrolyte balance in the wallabies *Thylogale thetis* and *Macropus eugenii* when given saline drinking water. *Comparative Biochemistry and Physiology*, **63A**, 135–9.

Hume, I. D., Rübsamen, K. & Engelhardt, W. v. (1980). Nitrogen metabolism and urea kinetics in the rock hyrax (*Procavia habessinica*). *Journal of Comparative Physiology*, **138**, 307–14.

Hume, I. D. & Warner, A. C. I. (1980). Evolution of microbial digestion in mammals. In *Digestive Physiology and Metabolism in Ruminants*, ed. Y. Ruckebusch & P. Thivend, pp. 615–34. Lancaster: MTP Press.

Hungate, R. E. (1966). *The Rumen and its Microbes*. New York: Academic Press.

Hungate, R. E., Phillips, G. D., McGregor, A., Hungate, D. P. & Buechner, H. K. (1959). Microbial fermentation in certain mammals. *Science*, **130**, 1192–4.

Hunsaker, D. II (1977). Ecology of the New World marsupials. In *The Biology of Marsupials*, ed. D. Hunsaker, II, pp. 95–156. New York: Academic Press.

Hutchinson, J. C. D. & Morris, S. (1936). The digestibility of dietary protein in the ruminant. I. Endogenous nitrogen excretion on a low nitrogen diet and in starvation. *Biochemical Journal*, **30**, 1682–94.

Inns, R. W. (1980). Ecology of the Kangaroo Island wallaby, *Macropus eugenii*

(Desmarest), in Flinders Chase National Park, Kangaroo Island. Ph.D. thesis, University of Adelaide, Adelaide.
International Committee on Veterinary Anatomical Nomenclature (1973). *Nomina Anatomica Veterinaria*, 2nd ed. Vienna: World Association of Veterinary Anatomists.
Irving, L., Krog, H. & Monson, M. (1955). The metabolism of some Alaskan animals in winter and summer. *Physiological Zoology*, **28**, 173–85.
Janis, C. (1976). The evolutionary strategy of the Equidae and the origins of rumen and cecal digestion. *Evolution*, **30**, 757–74.
Jarman, P. J. (1973). The social organization of antelope in relation to their ecology. *Behaviour*, **48**, 215–67.
Johnson, K. A. (1977). Ecology and management of the red-necked pademelon, *Thylogale thetis*, on the Dorrigo Plateau of northern New South Wales. Ph.D. thesis, University of New England, Armidale.
Johnson, K. (1980). Diet of the bilby, *Macrotis lagotis* in the western desert regions of central Australia. *Bulletin of the Australian Mammal Society*, **6**, 46–7 (Abstract).
Johnstone, J. (1898). On the gastric glands of the Marsupialia. *Journal of the Linnean Society*, **27**, 1–14.
Jung, H.-J. G. (1977). Responses of mammalian herbivores to secondary plant compounds. *The Biologist*, **59**, 123–36.
Kaethner, M. M. & Good, B. F. (1975). Seasonal thyroid activity in the tammar wallaby, *Macropus eugenii* (Desmarest). *Australian Journal of Zoology*, **23**, 363–9.
Kakulas, B. A. (1961). Myopathy affecting the Rottnest quokka (*Setonix brachyurus*) reversed by α-tocopherol. *Nature*, **191**, 402–3.
Kakulas, B. A. (1963a). Trace quantities of selenium ineffective in the prevention of nutitional myopathy in the Rottnest quokka (*Setonix brachyurus*). *Australian Journal of Science*, **25**, 313–14.
Kakulas, B. A. (1963b). Influence of the size of enclosure on the development of myopathy in the captive Rottnest quokka. *Nature*, **198**, 673–4.
Kakulas, B. A. (1966). Regeneration of skeletal muscle in the Rottnest quokka. *Australian Journal of Experimental Biology and Medical Science*, **44**, 673–88.
Kaldor, I. & Ezekiel, E. (1962). Iron content of mammalian breast milk; measurements in the rat and in a marsupial. *Nature*, **196**, 175.
Kaufmann, J. H. (1974). Habitat use and social organisation of nine sympatric species of macropodid marsupials. *Journal of Mammalogy*, **55**, 66–80.
Kay, G. I., Wheeler, H. O., Whitlock, R. T. & Lane, N. (1966). Fluid transport in the rabbit gallbladder. A combined physiological and electron microscope study. *Journal of Cell Biology*, **30**, 237–68.
Kempton, T. J. (1972). The efficiency of utilization of a roughage diet and the rate of passage of digesta in the grey kangaroo, *Macropus giganteus* (Shaw), and the Merino sheep. Unpublished thesis, University of New England, Armidale.
Kempton, T. J., Murray, A. M. & Leng, R. A. (1976). Rates of production of methane in the grey kangaroo and sheep. *Australian Journal of Biological Sciences*, **29**, 209–14.
Kennedy, P. M. & Heinsohn, G. E. (1974). Water metabolism of two marsupials – the brush-tailed possum, *Trichosurus vulpecula* and the rock-wallaby, *Petrogale inornata* in the wild. *Comparative Biochemistry and Physiology*, **47A**, 829–34.
Kennedy, P. M. & Hume, I. D. (1978). Recycling of urea nitrogen to the gut of the tammar wallaby (*Macropus eugenii*). *Comparative Biochemistry and Physiology*, **61A**, 117–21.
Kennedy, P. M. & Macfarlane, W. V. (1971). Oxygen consumption and water turnover of the fat-tailed marsupials *Dasycercus cristicauda* and *Sminthopsis crassicaudata*. *Comparative Biochemistry and Physiology*, **40A**, 723–32.

Kennedy, P. M. & Milligan, L. P. (1980). The degradation and utilization of endogenous urea in the gastrointestinal tract of ruminants: A review. *Canadian Journal of Animal Science*, **60**, 205–21.

Kerry, K. R. (1969). Intestinal disaccharidase activity in a monotreme and eight species of marsupials (with an added note on the dissacharidases of five species of sea birds). *Comparative Biochemistry and Physiology*, **29**, 1015–22.

Kinnear, A. & Shield, J. W. (1975). Metabolism and temperature regulation in marsupials. *Comparative Biochemistry and Physiology*, **52A**, 235–46.

Kinnear, J. E. (1970). Nitrogen metabolism of macropods with special reference to the tammar (*Macropus eugenii*). Ph.D. thesis, University of Western Australia, Perth.

Kinnear, J. E. & Brown, G. D. (1967). Minimum heart rates of marsupials. *Nature*, **215**, 1501.

Kinnear, J. E., Cockson, A., Christensen, P. & Main, A. R. (1979). The nutritional biology of the ruminants and ruminant-like mammals – a new approach. *Comparative Biochemistry and Physiology*, **64A**, 357–65.

Kinnear, J. E. & Main, A. R. (1975). The recycling of urea nitrogen by the wild tammar wallaby (*Macropus eugenii*) – a 'ruminant-like' marsupial. *Comparative Biochemistry and Physiology*, **51A**, 793–810.

Kinnear, J. E., Purohit, K. G. & Main, A. R. (1968). The ability of the tammar wallaby (*Macropus eugenii*, Marsupialia) to drink sea water. *Comparative Biochemistry and Physiology*, **25**, 761–82.

Kirkpatrick, T. H. (1965). Studies of Macropodidae in Queensland. I. Food preferences of the grey kangaroo (*Macropus major* Shaw). *Queensland Journal of Agricultural Science*, **22**, 89-93.

Kirsch, J. A. W. (1977a). The classification of marsupials. In *The Biology of Marsupials*, ed. D. Hunsaker, II, pp. 1–15. New York: Academic Press.

Kirsch, J. A. W. (1977b). Appendix: Notes on nomenclature. In *The Biology of Marsupials*, ed. D. Hunsaker, II, pp. 521–4. New York: Academic Press.

Kirsch, J. A. W. (1977c). The comparative serology of Marsupialia, and a classification of marsupials. *Australian Journal of Zoology, Supplementary Series*, **52**, 1–152.

Kirsch, J. A. W. & Calaby, J. H. (1977). The species of living marsupials: an annotated list. In *The Biology of Marsupials*, ed. B. Stonehouse & D. Gilmore, pp. 9–26. London: Macmillan.

Kirsch, J. A. W. & Waller, P. F. (1979). Notes on the trapping and behaviour of the Caenolistidae (Marsupialia). *Journal of Mammalogy*, **60**, 390–5.

Kleiber, M. (1932). Body size and metabolism. *Hilgardia*, **6**, 315–53.

Kleiber, M. (1961). *The Fire of Life*. New York: John Wiley.

Krause, W. J. (1970). Brunner's glands of the echidna. *Anatomical Record*, **167**, 473–88.

Krause, W. J. (1971). Brunner's glands of the duckbilled platypus (*Ornithorhynchus anatinus*). *American Journal of Anatomy*, **132**, 147–66.

Krause, W. J. (1972). The distribution of Brunner's glands in 55 marsupial species native to the Australian region. *Acta Anatomica*, **82**, 17–33.

Krause, W. J. & Leeson, C. R. (1973). The stomach gland patch of the koala (*Phascolarctos cinereus*). *Anatomical Record*, **176**, 475–88.

Lamprey, H. F. (1964). Estimation of the large mammal densities, biomass and energy exchange in the Tarangire Game Reserve and the Masai Steppe in Tanganyika. *East African Wildlife Journal*, **2**, 1–46.

Langer, P. (1979). Functional anatomy and ontogenetic development of the stomach in the macropodine species *Thylogale stigmatica* and *Thylogale thetis* (Mammalia: Marsupialia). *Zoomorphologie*, **93**, 137–51.

Langer, P. (1980a). Anatomy of the stomach in three species of Potoroinae (Marsupialia: Macropodidae). *Australian Journal of Zoology*, **28**, 19–31.

Langer, P. (1980b). Stomach evolution in the Macropodidae Owen, 1839 (Mammalia: Marsupialia). *Zeitschrift für Zoologische Systematik und Evolutionsforschung*, **18**, 211–32.

Langer, P., Dellow, D. W. & Hume, I. D. (1980). Stomach structure and function in three species of macropodine marsupials. *Australian Journal of Zoology*, **28**, 1–18.

Lay, D. W. (1942). Ecology of the opossum in eastern Texas. *Journal of Mammalogy*, **23**, 147–59.

Lee, A. K., Bradley, A. J. & Braithwaite, R. W. (1977). Corticosteroid levels and male mortality in *Antechinus stuartii*. In *The Biology of Marsupials*, ed. B. Stonehouse & D. Gilmore, pp. 209–20. London: Macmillan.

Leng, R. A. (1970). Glucose synthesis in ruminants. *Advances in Veterinary Science and Comparative Medicine*, **14**, 209–60.

Leng, R. A., Corbett, J. L. & Brett, D. J. (1968). Rates of production of volatile fatty acids in the rumen of grazing sheep and their relation to ruminal concentrations. *British Journal of Nutrition*, **22**, 57–68.

Leng, R. A. & Leonard, G. J. (1965). Measurement of the rates of production of acetic, propionic and butyric acids in the rumen of sheep. *British Journal of Nutrition*, **19**, 469–84.

Lifson, N. & McClintock, R. (1966). Theory of use of the turnover rates of body water for measuring energy and material balance. *Journal of Theoretical Biology*, **12**, 46–74.

Lintern, S. (1970). Aspects of nitrogen metabolism in the Kangaroo Island wallaby – *Protemnodon eugenii* (Desmarest). Ph.D. thesis, University of Adelaide, Adelaide.

Lintern, S. M. & Barker, S. (1969). Renal retention of urea in the Kangaroo Island wallaby, *Protemnodon eugenii* (Desmarest). *Australian Journal of Experimental Biology and Medical Science*, **47**, 243–50.

Lintern-Moore, S. (1973a). Incorporation of dietary nitrogen into microbial nitrogen in the forestomach of the Kangaroo Island wallaby *Protemnodon eugenii* (Desmarest). *Comparative Biochemistry and Physiology*, **44A**, 75–82.

Lintern-Moore, S. (1973b). Utilization of dietary urea by the Kangaroo Island wallaby *Protemnodon eugenii* (Desmarest). *Comparative Biochemistry and Physiology*, **46A**, 345–51.

Lönnberg, E. (1902). On some remarkable digestive adaptations in diprotodont marsupials. *Proceedings of the Zoological Society of London*, **73**, 12–31.

McClymont, G. L. (1968). Selectivity and intake in the grazing ruminant. In *Handbook of Physiology*, section 6, vol. 1, chapt. 9, pp. 129–37. Washington, DC: American Physiological Society.

McEwan, E. H. (1970). Energy metabolism of barren ground caribous (*Rangifer tarandus*). *Canadian Journal of Zoology*, **48**, 391–2.

Macfarlane, W. V. (1965). Water metabolism of desert ruminants. In *Studies in Physiology*, ed. D. R. Curtis & A. K. M. McIntyre, pp. 191–9. Berlin: Springer-Verlag.

Macfarlane, W. V. (1976), Ecophysiological hierarchies. *Israel Journal of Medical Sciences*, **12**, 723–31.

Macfarlane, W. V. (1977). The ecophysiology of water in desert organisms. In *Water, Planets, Plants and People*, ed. A. K. McIntyre, pp. 108–43. Canberra: Australian Academy of Science.

Macfarlane, W. V., Howard, B., Haines, H., Kennedy, P. M. & Sharp, C. M. (1971). Hierarchy of water and energy turnover of desert animals. *Nature*, **234**, 483–4.

Macfarlane, W. V., Morris, R. J. H., Howard, B., McDonald, J. & Budtz-Olsen, O. E. (1961). Water and electrolyte changes in tropical merino sheep exposed to dehydration during summer. *Australian Journal of Agricultural Research*, **12**, 889–912.

McIlroy, J. C. (1973). Aspects of the ecology of the common wombat, *Vombatus ursinus* (Shaw, 1800). Ph.D. thesis, Australian National University, Canberra.

McIntosh D. L. (1966). The digestibility of two roughages and the rates of passage of their residues by the red kangaroo, *Megaleia rufa* (Desmarest) and the merino sheep. *CSIRO Wildlife Research*, **11**, 125–35.

McKay, G. M. (1980). Nomenclature of the gliding phalanger genera *Petaurus* and *Petauroides*. *Bulletin of the Australian Mammal Society*, **6**, 51 (Abstract).

McKenzie, R. A. (1978). The caecum of the koala, *Phascolarctos cinereus*. Light scanning and transmission electron microscopic observations on its epithelium and flora. *Australian Journal of Zoology*, **26**, 249–56.

Mackenzie, W. C. (1918). *The Gastro-Intestinal Tract in Monotremes and Marsupials*. Melbourne: Critchley Parker.

McLeod, M. N. (1974). Plant tannins – their role in forage quality. *Nutrition Abstracts and Reviews*, **44**, 803–15.

MacMillen, R. E. & Lee, A. K. (1969). Water metabolism of Australian hopping mice. *Comparative Biochemistry and Physiology*, **28**, 493–514.

MacMillen, R. E. & Nelson, J. E. (1969). Bioenergetics and body size in dasyurid marsupials. *American Journal of Physiology*, **217**, 1246–51.

McNab, B. K. (1969). The economics of temperature regulation in neotropical bats. *Comparative Biochemistry and Physiology*, **31**, 227–68.

McNab, B. K. (1978). The comparative energetics of neotropical marsupials. *Journal of Comparative Physiology*, **125**, 115–28.

McNab, B. K. & Morrison, P. R. (1963). Body temperature and metabolism in subspecies of *Peromyscus* from arid and mesic environments. *Ecological Monographs*, **33**, 63–82.

Main, A. R. (1970). Measures of wellbeing in populations of herbivorous macropod marsupials. In *Dynamics of Populations*, ed. P. J. den Boer & G. R. Gradwell, pp. 159–73. Wageningen: Centre for Agricultural Publishing and Documentation.

Main, A. R. & Yadav, M. (1971). Conservation of macropods in reserves in Western Australia. *Biological Conservation*, **3**, 123–33.

Maller, O., Clark, J. M. & Kare, M. R. (1965). Short term caloric regulations in the adult opossum (*Didelphis virginiana*). *Proceedings of the Society for Experimental Biology and Medicine*, **118**, 275–7.

Marples, T. J. (1973). Studies on the marsupial glider, *Schoinobates volans* (Kerr). IV. Feeding biology. *Australian Journal of Zoology*, **21**, 213–16.

Marshall, L. G. (1972). Evolution of the peramelid tarsus. *Proceedings of the Royal Society of Victoria*, **8**, 51–60.

Martin, C. J. (1903). Thermal adjustment and respiratory exchange in monotremes and marsupials – a study in the development of homoeothermism. *Philosophical Transactions of the Royal Society of London, Series B*, **195**, 1–37.

Maxwell, G. M., Elliott, R. B. & Kneebone, G. M. (1964). Hemodynamics of kangaroos and wallabies. *American Journal of Physiology*, **206**, 967–70.

Maynard, L. A. & Loosli, J. K. (1962). *Animal Nutrition*, 5th ed. New York: McGraw-Hill.

Maynes, G. M. (1974). Occurrence and field recognition of *Macropus parma*. *Australian Zoologist*, **18**, 72–87.

Mead, R. J., Oliver, A. J. & King, D. R. (1979). Metabolism and defluorination of fluoroacetate in the brush-tailed possum (*Trichosurus vulpecula*). *Australian Journal of Biological Sciences*, **32**, 15–26.

Mercer, J. R. & Annison, E. F. (1976). Utilization of nitrogen in ruminants. In *Protein Metabolism and Nutrition*, ed. D. J. A. Cole, K. N. Boorman, P. J. Buttery, D. Lewis, R. J. Neale & H. Swan, pp. 397–416. London: Butterworths.

Miller, T. & Bradshaw, S. D. (1979). Adrenocortical function in a field population of a macropodid marsupial (*Setonix brachyurus*, Quoy & Gaimard). *Journal of Endocrinology*, **82**, 152–70.

Minchin, A. K. (1937). Notes on the weaning of a young koala (*Phascolarctos cinereus*). *Records of the South Australian Museum*, **6**, 1–6.

Mitchell, H. H. (1962). *Comparative Nutrition of Man and Domestic Animals*, vol. 1. New York: Academic Press.

Mitchell, P. C. (1905). On the intestinal tract of mammals. *Transactions of the Zoological Society of London*, **17**, 437–537.

Mitchell, P. C. (1916). Further observations on the intestinal tracts of mammals. *Proceedings of the Zoological Society of London*, 1916, 183–251.

Moir, R. J. (1965). The comparative physiology of ruminant-like animals. In *Physiology of Digestion in the Ruminant*, ed. R. W. Dougherty, pp. 1–14. London: Butterworths.

Moir, R. J. (1968). Ruminant digestion and evolution. In *Handbook of Physiology*, section 6 (Alimentary canal), vol. 5 (Bile; Digestion; Ruminal Physiology), 2673–94. Washington, DC: American Physiological Society.

Moir, R. J., Somers, M. & Waring, H. (1956). Studies on marsupial nutrition. I. Ruminant-like digestion in a herbivorous marsupial *Setonix brachyurus* (Quoy and Gaimard). *Australian Journal of Biological Sciences*, **9**, 293–304.

Moir, R. J. & Williams, V. J. (1950). Ruminal flora studies in sheep. II. The effect of the levels of nitrogen intake upon the total number of free micro-organisms in the rumen. *Australian Journal of Scientific Research*, **B3**, 381–92.

Mollison, B. C. (1960). Food regurgitation in Bennett's wallaby, *Protemnodon rufogrisea* (Desmarest) and the scrub wallaby, *Thylogale billardieri* (Desmarest). *CSIRO Wildlife Research*, **5**, 87–8.

Morrison, P. R. (1946). Temperature regulation in three central American mammals. *Journal of Cellular and Comparative Physiology*, **27**, 125–37.

Morrison, P. R. (1965). Body temperatures in some Australian mammals. IV. Dasyuridae. *Australian Journal of Zoology*, **13**, 173–87.

Morton, S. R. (1980a). An ecological study of *Sminthopsis crassicaudata* (Marsupialia: Dasyuridae). II. Behaviour and social organization. *Australian Wildlife Research*, **5**, 163–82.

Morton, S. R. (1980b). An ecological study of *Sminthopsis crassicaudata* (Marsupialia: Dasyuridae). III. Reproduction and life history. *Australian Wildlife Research*, **5**, 183–211.

Morton, S. R. (1980c). Field and laboratory studies of water metabolism in *Sminthopsis crassicaudata* (Marsupialia: Dasyuridae). *Australian Journal of Zoology*, **28**, 213–27.

Morton, S. R. (1980d). Ecological correlates of caudal fat storage in small mammals. *Australian Mammalogy*, **3**, 81–6.

Myers, K., Bailey, P. J. & Dudzinski, M. L. (1976). The effects of a severe drought on the morphology of the adrenal glands and other indices of health in the red kangaroo, *Megaleia rufa* (Desmarest) in north-western New South Wales. *Australian Journal of Ecology*, **1**, 289–302.

Nagy, J. G., Steinhoff, H. W. & Ward, G. M. (1964). Effects of essential oils of sagebrush on deer rumen microbial function. *Journal of Wildlife Management*, **28**, 785–90.

Nagy, K. A. & Milton, K. (1979). Energy metabolism and food consumption by wild howler monkeys (*Alouatta palliata*). *Ecology*, **60**, 475–80.

Nagy, K. A., Seymour, R. S., Lee, A. K. & Braithwaite, R. (1979). Energy and water budgets in free-living *Antechinus stuartii* (Marsupialia: Dasyuridae). *Journal of Mammalogy*, **59**, 60–8.

Nagy, K. A., Shoemaker, V. H. & Costa, W. R. (1976). Water, electrolyte and nitrogen budgets of jack-rabbits (*Lepus californicus*) in the Mojave Desert. *Physiological Zoology*, **49**, 351–63.

Nakajima, Y., Shantha, T. R. & Bourne, G. H. (1969). Histochemical detection of L-gulonolactone: phenazine methosulfate oxidoreductase activity in several mammals with special reference to synthesis of vitamin C in primates. *Histochemie*, **18**, 293–301.

Nestel, P. J., Havenstein, N., Whyte, H. M., Scott, T. W. & Cook, L. J. (1973). Lowering of plasma cholesterol and enhanced sterol excretion with the consumption of polyunsaturated ruminant fats. *New England Journal of Medicine*, **228**, 279–82.

Newsome, A. E. (1964). Anoestrus in the red kangaroo, *Megaleia rufa* (Desmarest). *Australian Journal of Zoology*, **12**, 9–17.

Newsome, A. E. (1965). The distribution of red kangaroos, *Megaleia rufa* (Desmarest), about sources of persistent food and water in central Australia. *Australian Journal of Zoology*, **13**, 289–99.

Newsome, A. E. (1966). The influence of food on breeding in the red kangaroo in central Australia. *CSIRO Wildlife Research*, **11**, 187–96.

Newsome, A. E. (1971). The ecology of red kangaroos. *Australian Zoologist*, **16**, 32–50.

Newsome, A. E. (1975). An ecological comparison of the two arid-zone kangaroos of Australia, and their anomalous prosperity since the introduction of ruminant stock to their environment. *Quarterly Review of Biology*, **50**, 389–424.

Nicol, S. C. (1976). Oxygen consumption and nitrogen metabolism in the potoroo, *Potorous tridactylus*. *Comparative Biochemistry and Physiology*, **55A**, 215–18.

Nicol, S. C. (1978a). Metabolism and temperature regulation in marsupials with particular reference to the potoroo, *Potorous tridactylus apicalis* Gould. Ph.D. thesis, University of Tasmania, Hobart.

Nicol, S. C. (1978b). Rates of water turnover in marsupials and eutherians: A comparative review with new data on the Tasmanian devil. *Australian Journal of Zoology*, **26**, 465–73.

Nolan, J. V. & Leng, R. A. (1972). Dynamic aspects of ammonia and urea metabolism in sheep. *British Journal of Nutrition*, **27**, 177–94.

Nolan, J. V. & Stachiw, S. (1979). Fermentation and nitrogen dynamics in Merino sheep given a low-quality-roughage diet. *British Journal of Nutrition*, **42**, 63–80.

Oh, H. K., Sakai, T., Jones, M. B. & Longhurst, W. M. (1967). Effect of various essential oils isolated from Douglas fir needles upon sheep and deer rumen microbial activity. *Applied Microbiology*, **15**, 777–84.

Oppel, A. (1896). *Lehrbuch der Vergleichenden Mikroskopischen Anatomie der Wirbeltiere*, vol. 1, *Der Magen*, pp. 286–98. Jena: Gustav Fischer.

Ørskov, E. R., Benzie, D. & Kay, R. N. B. (1970). The effects of feeding procedure on closure of the oesophageal groove in young sheep. *British Journal of Nutrition*, **24**, 785–95.

Osgood, W. H. (1921). A monographic study of the American marsupial, *Caenolestes*. *Field Museum of Natural History, Zoological Series*, **14**, 1–162.

Owaga, M. L. (1975). The feeding ecology of wildebeest and zebra in Athi-Kaputei Plains. *East African Wildlife Journal*, **13**, 375–83.

Owen, R. (1834). Notes on the anatomy of a new species of kangaroo (*Macropus parryi*, Benn.). *Proceedings of the Zoological Society of London*, Part II, 152.

Owen, R. (1839–47). Marsupialia. In *The Cyclopaedia of Anatomy and Physiology*, ed. R. B. Todd, vol. 3. London: Longman, Brown, Green, Longmans & Roberts.

Owen, R. (1868). *On the Anatomy of Vertebrates*, vol. III, *Mammals*, pp. 411–20. London: Longmans, Green & Co.

Owen, W. H. & Thomson, J. A. (1965). Notes on the comparative ecology of the common brush-tail and mountain possums in eastern Victoria. *Victorian Naturalist*, **82**, 216–17.

Parker, D. S. & McMillan, R. T. (1976). The determination of volatile fatty acids in the caecum of the conscious rabbit. *British Journal of Nutrition*, **35**, 365–71.

Parsons, F. G. (1903). On the anatomy of the pig-footed bandicoot. *Journal of the Linnean Society, Zoology*, **29**, 64–80.

Paton, D. (1979). Ecology of the New World Honeyeater. Ph.D. thesis, Monash University, Melbourne.

Pernetta, J. C. (1976). Diets of the shrews *Sorex araneus* L. and *Sorex minutus* L. in Wytham grassland. *Journal of Animal Ecology*, **45**, 899–912.

Petter, J. J., Schilling, A. & Pariente, G. (1971). Observations écoéthologiques sur deux lémuriens malgaches nocturnes: *Phaner furcifer et Microcebus coquereli*. *Terre Vie*, **25**, 287–327.

Pickard, D. W. & Stevens, C. E. (1972). Digesta flow through the rabbit's large intestine. *American Journal of Physiology*, **222**, 1161–6.

Plakke, R. K. & Pfeiffer, E. W. (1965). Influence of plasma urea on urine concentration in the opossum (*Didelphis marsupialis virginiana*), *Nature*, **207**, 866–7.

Plakke, R. K. & Pfeiffer, E. W. (1970). Urea, electrolyte and total solute excretion following water deprivation in the opossum (*Didelphis marsupialis virginiana*). *Comparative Biochemistry and Physiology*, **34**, 325–32.

Poole, W. E. (1979). The status of endangered Australian Macropodidae. In *The Status of Endangered Australasian Wildlife*, ed. M. J. Tyler, pp. 13–27. Adelaide: Royal Zoological Society of South Australia.

Prince, R. I. T. (1976). Comparative studies of aspects of nutritional and related physiology in macropod marsupials. Ph.D. thesis, University of Western Australia, Perth.

Prins, R. A. (1977). Biochemical activities of gut micro-organisms. In *Microbial Ecology of the Gut*, ed. R. T. J. Clarke & T. Bauchop, pp. 73–183. New York: Academic Press.

Prior, R. L., Hintz, H. F., Lowe, J. E. & Visek, W. J. (1974). Urea recycling and metabolism of ponies. *Journal of Animal Science*, **38**, 565–71.

Purohit, K. G. (1971). Absolute duration of survival of tammar wallaby (*Macropus eugenii*, Marsupialia) on sea water and dry food. *Comparative Biochemistry and Physiology*, **39A**, 473–81.

Ramsay, B. A. (1966). Field nutrition in the Rottnest quokka. Master of Science thesis, University of Western Australia, Perth.

Raven, H. C. & Gregory, W. K. (1946). Adaptive branching of the kangaroo family in relation to habitat. *American Museum Novitates*, **1309**, 1–33.

Redgrave, T. G. & Vickery, D. M. (1973). The polyunsaturated nature of horse and kangaroo fats. *Medical Journal of Australia*, **2**, 1116–18.

Reid, I. A. (1977). Some aspects of renal physiology in the Brush-tailed possum, *Trichosurus vulpecula*. In *The Biology of Marsupials*, ed. B. Stonehouse & D. Gilmore, pp. 393–410. London: Macmillan.

Reynolds, H. C. (1945). Some aspects of the life history and ecology of the opossum in central Missouri. *Journal of Mammalogy*, **26**, 361–79.

Rhoades, D. F. & Cates, R. G. (1976). Toward a general theory of plant antiherbivore chemistry. *Recent Advances in Phytochemistry*, **10**, 168–213.

Richardson, B. J. & McDermid, E. M. (1978). A comparison of genetic relationships within the Macropodidae as determined from allozyme, cytological and immunological data. *Australian Mammalogy*, **2**, 43–51.

Richardson, B. J. & Sharman, G. B. (1976). Biochemical and morphological observations on the wallaroos (Macropodidae: Marsupialia) with a suggested new taxonomy. *Journal of Zoology, London*, **179**, 499–513.

Richardson, K. C. (1980). The structure and radiographic anatomy of the alimentary tract of the tammar wallaby, *Macropus eugenii* (Marsupialia). I. The stomach. *Australian Journal of Zoology*, **28**, 367–79.

Richardson, K. C. & Wyburn, R. S. (1980). The structure and radiographic analysis of the alimentary tract of the tammar wallaby, *Macropus eugenii* (Marsupialia). II. The intestines. *Australian Journal of Zoology*, **28**, 367–79.

Richmond, C. R., Langham, W. H. & Trujillo, T. T. (1962). Comparative metabolism of tritiated water by mammals. *Journal of Cellular and Comparative Physiology*, **59**, 45–53.

Ride, W. D. L. (1964). A review of Australian fossil marsupials. *Journal of the Royal Society of Western Australia*, **47**, 97–131.

Ride, W, D. L. (1970). *A Guide to the Native Animals of Australia*. Melbourne: Oxford University Press.

Ride, W. D. L. & Tyndale-Biscoe, C. H. (1962). Mammals. In *The Results of an Expedition to Bernier and Dorre Islands, Shark Bay, Western Australia*, ed. A. J. Fraser, pp. 54–97, Fauna Bulletin No. 2, Fisheries Department of Western Australia.

Robinson, K. W. (1954). Heat tolerances of Australian monotremes and marsupials. *Australian Journal of Biological Sciences*, **7**, 348–60.

Robinson, K. W. & Morrison, P. R. (1957). The reaction to hot atmospheres of various species of Australian marsupial and placental mammals. *Journal of Cellular and Comparative Physiology*, **49**, 455–78.

Rogers, Q. R., Morris, J. G. & Freedland, R. A. (1977). Lack of hepatic enzymatic adaptation to low and high levels of dietary protein in the adult cat. *Enzyme*, **22**, 348–56.

Rosenmann, M. & Morrison, P. (1963). Physiological response to heat and dehydration in the guanaco. *Physiological Zoology*, **36**, 45–51.

Rübsamen, K., Heller, R., Lawrenz, H. & Engelhardt, W. v. (1979a). Water and energy metabolism in the rock hyrax (*Procavia habessinica*). *Journal of Comparative Physiology*, **131**, 303–9.

Rübsamen, K., Nolda, V. & Engelhardt, W. v. (1976). Difference in the specific activity of tritium labelled water in blood, urine and evaporative water in rabbits. *Comparative Biochemistry and Physiology*, **62A**, 279–82.

Russell, E. M. (1974). The biology of kangaroos (Marsupialia-Macropodidae). *Mammal Reviews*, **4**, 1–59.

Sampson, J. C. (1971). The biology of *Bettongia penicillata* Gray, 1837. Ph.D. thesis, University of Western Australia, Perth.

Sanson, G. D. (1976). The evolution of mastication in the Macropodinae. *Bulletin of the Australian Mammal Society*, **3**, 61–2 (Abstract).

Sanson, G. D. (1978). The evolution and significance of mastication in the Macropodidae. *Australian Mammalogy*, **2**, 23–8.

Sanson, G. D. (1980). The morphology and occlusion of the molariform check teeth in some Macropodinae (Marsupialia: Macropodidae). *Australian Journal of Zoology*, **28**, 341–65.

Sanson, G. D. & Miller, W. A. (1979). Mechanism of molar progression in macropods. *Anatomical Record*, **193**, 674 (Abstract).

Schäfer, E. A. & Williams, D. J. (1876). On the structure of the mucous membrane of the stomach in the kangaroos. *Proceedings of the Zoological Society of London*, No. XII, 165–77.

Schmidt-Nielsen, B. (1958). Urea excretion in mammals. *Physiological Reviews*, **38**, 139–68.

Schmidt-Nielsen, B. & O'Dell, R. (1959). Effect of diet on distribution of urea and electroytes in kidneys of sheep. *American Journal of Physiology*, **197**, 856–60.

Schmidt-Nielsen, B. & O'Dell, R. (1961). Structure and concentrating mechanism in the mammalian kidney. *American Journal of Physiology*, **200**, 1119–24.

Schmidt-Nielsen, B., O'Dell, R. & Osaki, H. (1961). Interdependence of urea and

electrolytes in production of a concentrated urine. *American Journal of Physiology*, **200**, 1125–32.
Schmidt-Nielsen, B. & Schmidt-Nielsen, K. (1950). Do kangaroo rats thrive when drinking sea water? *American Journal of Physiology*, **160**, 291–4.
Schmidt-Nielsen, B., Schmidt-Nielsen, K., Houpt, T. R. & Jarnum, S. A. (1957). Urea excretion in the camel. *American Journal of Physiology*, **188**, 477–83.
Schmidt-Nielsen, K. (1964). *Desert Animals*. Oxford: Clarendon Press.
Schmidt-Nielsen, K., Crawford, E. C., Newsome, A. E., Rawson, K. S. & Hammel, H. T. (1967). Metabolic rate of camels: effects of body temperature and dehydration. *American Journal of Physiology*, **212**, 341–6.
Schmidt-Nielsen, K. & Newsome, A. E. (1962). Water balance in the mulgara (*Dasycercus cristicauda*), a carnivorous desert marsupial. *Australian Journal of Biological Sciences*, **15**, 683–9.
Scholander, P. F., Hock, R., Walters, V., Johnson, F. & Irving, L. (1950). Heat regulation in some arctic and tropical mammals and birds. *Biological Bulletin*, **99**, 237–58.
Schultz, W. (1976). Magen-Darm-Kanal der Monotremen und Marsupialier. *Handbuch der Zoologie*, **8(53)**, 1–177.
Scott, I. M., Yousef, M. K. & Johnson, H. D. (1976). Plasma thyroxine levels of mammals: desert and mountain. *Life Sciences*, **19**, 807–12.
Setchell, P. J. (1974). Studies of the control of thyroid function in marsupials. Ph.D. thesis, University of Adelaide, Adelaide.
Sharman, G. B. (1959). Marsupial reproduction. *Monographiae Biologicae*, **8**, 332–68.
Sharman, G. B. & Calaby, J. H. (1964). Reproductive behaviour in the red kangaroo, *Megaleia rufa*, in captivity. *CSIRO Wildlife Research*, **9**, 58–85.
Shield, J. W. (1959). Rottnest field studies concerned with the quokka. *Journal of the Royal Society of Western Australia*, **42**, 76–82.
Shield, J. (1965). A breeding season difference in two populations of the Australian macropod marsupial (*Setonix brachyurus*). *Journal of Mammalogy*, **45**, 616–25.
Shkolnik, A. & Schmidt-Nielsen, K. (1976). Temperature regulation in hedgehogs from temperate and desert environments. *Physiological Zoology*, **49**, 56–64.
Sibbald, I. R., Sinclair, D. G., Evans, E. V. & Smith, D. L. T. (1962). The rate of passage of feed through the digestive tract of the mink. *Canadian Journal of Biochemistry and Physiology*, **40**, 1391–4.
Slater, J. & Jones, R. J. (1971). Estimation of diets selected by grazing animals from microscopic analysis of the faeces – a warning. *Journal of the Australian Institute of Agricultural Science*, **37**, 238—40.
Smith, A. (1978). Feeding strategies of the sugar glider in south Gippsland. *Bulletin of the Australian Mammal Society*, **5**, 17 (Abstract).
Smith, A. (1980). The diet and ecology of Leadbeater's possum and the sugar glider. Ph.D. thesis, Monash University, Melbourne.
Smith, A. & Russell, R. (1982). Diet of the Yellow-bellied Glider, *Petaurus australis*. *Australian Mammalogy*, **5**, (in press).
Smith, M. J. (1973). *Petaurus breviceps*. *Mammalian Species*, No. 30, 1–5.
Smuts, D. B. (1935). The relation between the basal metabolism and the endogenous nitrogen metabolism with particular reference to the estimation of the maintenance requirement of protein. *Journal of Nutrition*, **9**, 403–33.
Smuts, D. B. & Marais, J. S. C. (1938). The endogenous nitrogen metabolism of sheep with special reference to the maintenance requirement of protein. *Onderstepoort Journal of Veterinary Science*, **11**, 131–9.
Sonntag, C. F. (1921). Contributions to the visceral anatomy and myology of the Marsupialia. *Proceedings of the Zoological Society of London*, **2**, 851–82.

Southwell, I. A. (1973). Variation in the leaf oil of *Eucalyptus punctata*. *Phytochemistry*, **12**, 1341–3.
Southwell, I. A. (1978). Essential oil content of koala food trees. In *The Koala*, ed. T. J. Bergin, pp. 62–74. Sydney: Zoological Parks Board of New South Wales.
Southwell, I. A., Flynn, T. M. & Degabriele, R. (1980). Metabolism of α- and β-pinene, p-cymene and 1,8-cineole in the brushtail possum, *Trichosurus vulpecula*. *Xenobiotica*, **10**, 17–23.
Sperber, I. (1944). Studies on the mammalian kidney. *Zoologiska Bidrag Fran Uppsala*, **22**, 249–431.
Sperber, I. (1968). Physiological mechanisms in herbivores for retention and utilization of nitrogenous compounds. In *Isotope Studies on the Nitrogen Chain*, pp. 209–19. Vienna: International Atomic Energy Agency.
Stanley, R. G. & Linskins, H. F. (1974). *Pollen*. Berlin: Springer-Verlag.
Stewart, C. M., Melvin, J. F., Ditchurne, N., Than, S. M. & Zerdoner, E. (1973). The effect of season of growth on the chemical composition of cambial saps of *Eucalyptus regnens* trees. *Oecologia*, **12**, 349–72.
Storr, G. M. (1961). Microscopic analysis of faeces, a technique for ascertaining the diet of herbivorous mammals. *Australian Journal of Biological Sciences*, **14**, 157–64.
Storr, G. M. (1963). Estimation of dry-matter intake in wild herbivores. *Nature*, **197**, 307–8.
Storr, G. M. (1964a). Studies on marsupial nutrition. 4. Diet of the quokka, *Setonix brachyurus* (Quoy and Gaimard) on Rottnest Island, Western Australia. *Australian Journal of Biological Sciences*, **17**, 469–81.
Storr, G. M. (1946b). The environment of the quokka (*Setonix brachyurus*) in the Darling Range, Western Australia. *Journal of the Royal Society of Western Australia*, **47**, 1–2.
Storr, G. M. (1968). Diet of kangaroos (*Megaleia rufa* and *Macropus robustus*) and merino sheep near Port Hedland, Western Australia. *Journal of the Royal Society of Western Australia*, **51**, 25–32.
Strahan, R. (1980). Recommended common names of Australian mammals. *Australian Mammal Society Bulletin*, **6**, 13–23.
Sutherland, A. (1897). The temperatures of reptiles, monotremes and marsupials. *Proceedings of the Royal Society of Victoria*, **9**, 57–67.
Swain, T. (1978). Phenolics in the environment. *Recent Advances in Phytochemistry*, **12**, 617–40.
Tan, T. M., Weston, R. H. & Hogan, J. P. (1971). Use of [103]Ru-labelled tris (1,10-phenanthroline) ruthenium(II) chloride as a marker in digestion studies with sheep. *International Journal of Applied Radiation and Isotopes*, **22**, 301–8.
Taylor, C. R. & Sale, J. B. (1969). Temperature regulation in the hyrax. *Comparative Biochemistry and Physiology*, **31**, 903–7.
Than, K. A. & McDonald, I. E. (1973). Adrenocortical function in the Australian Brush-tailed Possum *Trichosurus vulpecula* (Kerr). *Journal of Endocrinology*, **58**, 97–109.
Thomson, J. A. & Owen, W. H. (1964). A field study of the Australian ringtail possum *Pseudocheirus peregrinus* (Marsupialia: Phalangeridae). *Ecological Monographs*, **34**, 27–52.
Tribe, D. E. & Peel, L. (1963). Body composition of the kangaroo (*Macropus* sp.). *Australian Journal of Zoology*, **11**, 273–89.
Troughton, E. (1965). *Furred Animals of Australia*, 8th ed. Sydney: Angus & Robertson.
Tyndale-Biscoe, H. (1973). *Life of Marsupials*. London: Edward Arnold.
Tyndale-Biscoe, C. H. & Calby, J. H. (1975). Eucalypt forests as refuge for wildlife. *Australian Forestry*, **38**, 117–33.
Tyndale-Biscoe, C. H. & Smith, R. F. C. (1969a). Studies on the marsupial glider

Schoinobates volans (Kerr). 2. Population, structure and regulatory mechanisms. *Journal of Animal Ecology*, **38**, 637–50.

Tyndale-Biscoe, C. H. & Smith, R. F. C. (1969b). Studies on the marsupial glider *Schoinobates volans* (Kerr). 3. Response to habitat destruction. *Journal of Animal Ecology*, **38**, 651–9.

Ullrey, D. E., Robinson, P. T. & Whetter, P. A. (1981). Composition of preferred and rejected *Eucalyptus* browse offered to captive Koalas, *Phascolarctos cinereus*. *Australian Journal of Zoology* (in press).

Underwood, E. J. (1977). *Trace Elements in Human and Animal Nutrition*, 4th ed. New York: Academic Press.

Van Soest, P. J. & Wine, R. H. (1967). Use of detergents in the analysis of fibrous feeds. IV. Determination of plant cell-wall constituents. *Journal of the Association of Official Agricultural Chemists*, **50**, 50–5.

Vogtsberger, L. M. & Barrett, G. W. (1973). Bioenergetics of captive red foxes. *Journal of Wildlife Management*, **37**, 495–500.

Vose, H. M. (1973). Feeding habits of the Western Australian Honey Possum, *Tarsipes spencerae*. *Journal of Mammalogy*, **54**, 245–7.

Wake, J. (1980). The field nutrition of the Rottnest Island quokka. Ph.D. thesis, University of Western Australia, Perth.

Wakefield, N. A. (1970). Notes on the glider-possum, *Petaurus australis* (Phalangeridae, Marsupialia). *Victorian Naturalist*, **87**, 221–36.

Walker, E. P. (1975). *Mammals of the World*, 3rd ed. Baltimore: Johns Hopkins University Press.

Waring, H., Moir, R. J. & Tyndale-Biscoe, C. H. (1966). Comparative physiology of marsupials. *Advances in Comparative Physiology and Biochemistry* **2**, 237–376.

Warner, A. C. I. (1981). The mean retention times of digesta markers in the gut of the tammar, *Macropus eugenii*. *Australian Journal of Zoology* (in press).

Weiss, M. & McDonald, I. R. (1966). Corticoid secretion in the Australian phalanger (*Trichosurus vulpecula*). *General and Comparative Endocrinology*, **7**, 345–51.

Wellard, G. A. & Hume, I. D. (1981a). Nitrogen metabolism and nitrogen requirements of the brushtail possum, *Trichosurus vulpecula* (Kerr). *Australian Journal of Zoology*, **29**, 147–56.

Wellard, G. A. & Hume, I. D. (1981b). Digestion and digesta passage in the brushtail possum, *Trichosurus vulpecula* (Kerr). *Australian Journal of Zoology*, **29**, 157–66.

Wells, R. T. (1968). Some aspects of the environmental physiology of the hairy-nosed wombat *Lasiorhinus latifrons* (Owen). Unpublished thesis, University of Adelaide, Adelaide.

Wells, R. T. (1973). Physiological and behavioural adaptations of the hairy-nosed wombat (*Lasiorhinus latifrons* Owen) to its arid environment. Ph.D. thesis, University of Adelaide, Adelaide.

Wells, R. T. (1978). Thermoregulation and acrivity rhythms in the hairy-nosed wombat, *Lasiorhinus latifrons* (Owen), (Vombatidae). *Australian Journal of Zoology*, **26**, 639–51.

West, G. C. (1960). Seasonal variation in the energy balance of the tree sparrow in relation to migration. *Auk*, **77**, 306–29.

Whitelaw, F. G., Hyldgaard-Jensen, J., Reid, R. S. & Kay, M. G. (1970). Volatile fatty acid production in the rumen of cattle given an all-concentrate diet. *British Journal of Nutrition*, **24**, 179–95.

Wilckens, M. (1872). *Utersuchungen über den Magen der wiederkauenden Hausthiere*. Berlin: von Wiegardt & Hempel.

Wilkinson, P. (1979). Urinary creatinine excretion by the macropod marsupials Red-necked Pademelon (*Thylogale thetis*) and Tammar Wallaby (*Macropus eugenii*). Unpublished thesis, University of New England, Armidale.

Wolf, L. L. & Hainsworth, F. R. (1971). Time and energy budgets of territorial hummingbirds. *Ecology*, **52**, 980–8.

Wood, D. H. (1970). An ecological study of *Antechinus stuartii* (Marsupialia) in a south-east Queensland rain forest. *Australian Journal of Zoology*, **18**, 185–207.

Wood, J. E. (1954). Food habits of furbearers of the upland Post Oak region in Texas. *Journal of Mammalogy*, **35**, 406–15.

Wood Jones, F. (1924). *The Mammals of South Australia*. Part II. *The Bandicoots and the Herbivorous Marsupials*. Adelaide: R. E. E. Rogers, Government Printer.

Woollard, P. (1971). Differential mortality of *Antechinus stuartii* (Macleay): nitrogen balance and somatic changes. *Australian Journal of Zoology*, **19**, 347–53.

Woolley, P. (1966). Reproduction in *Antechinus* spp. and other dasyurid marsupials. In *Comparative Biology of Reproduction in Mammals*, ed. I. W. Rowlands, *Symposia of the Zoological Society of London*, vol. 15, pp. 281–94.

Yadav, M. (1979). The kidney types of some Western Australian macropod marsupials. *Mammalia*, **43**, 225–33.

Yousef, M. K., Dill, D. B. & Mayes, M. G. (1970). Shifts in body fluids during dehydration in the burro, *Equis asinus*. *Journal of Applied Physiology*, **29**, 345–9.

Yousef, M. K., Johnson, H. D., Bradley, W. G. & Seif, S. M. (1974). Tritiated water turnover rate in rodents: Desert and mountain. *Physiological Zoology*, **47**, 153–62.

Index

Abrolhos Islands 184–8
Acacia
  gum as food 65, 67, 204
  seeds as food 190
Acrobates pygmaeus 62, 64–5
Adrenal function 155, 177, 220
Aepyprymnus rufescens 200, 202–6
Ammonia in macropodine stomach 142–3, 145–6, 155
Amylase activity 120
Anaemia 48, 182
Antechinomys laniger 4, 38
Antechinus bellus 48
Antechinus bilarni 48
Antechinus flavipes 48
Antechinus macdonnellensis 4, 6, 7
Antechinus minimus 48
Antechinus stuartii 4–7, 15, 27, 30–2, 40–1, 46–9, 54–5, 222
Antechinus swainsonii 15, 32, 40–1, 48, 56, 108
Ascorbic acid biosynthesis 221–5
  bandicoots 222, 224
  marsupials 222–5
  monotremes 222–4

Bacteria 85, 140
Basal metabolic rate 1–3
  aquatic carnivores 6, 7, 11
  arboreal folivores 5, 7, 11, 86, 88
  arboreal omnivores 5, 11, 15–16
  bandicoots 5–7
  dasyurids 4–7, 13, 25, 44
  didelphids 6, 7, 11
  eutherians 3–5, 8–12, 14–17, 25
  macropodids 5, 20
  terrestrial carnivores 11
  terrestrial grazer-browsers 11
  terrestrial omnivore-insectivores 11
  wombats 6, 7
Bernier Island 51, 167
Bettongia lesueur 178, 200–2, 205
Bettongia penicillata 178, 200–8
Burramys parvus 50, 62

Caecectomy, effects in Brushtails 79–80
Caecum
  arboreal folivores 77, 80, 84, 89, 99
  arboreal omnivores 65–7
  bandicoots 52–4
  caenolestids 35–6
  dasyurids 38–9
  didelphids 37, 57
  macropodids 115, 147, 206
  wombats 71–3
Caenolestes obscurus 29, 35, 36–7, 39
Carbohydrate metabolism, macropodines 151–5
Cardiac glandular mucosa 137–8
Cardio-gastric gland 35, 71, 85, 204, 206
Carnivore diets 27–30
Cercartetus nanus 50, 62, 222
Chaeropus castanotis 51–2, 54
Chironectes minimus 6, 7, 11, 29, 36–7
Cobalt 215
Colon
  arboreal folivores 77, 80, 84, 89, 99
  arboreal omnivores 65–7
  caenolestids 35–6
  dasyurids 38
  didelphids 37, 57
  macropodids 115, 147, 206
  wombats 71–3
Condition factor, Quokkas 180–3
Copper 213–14
Coprophagy 102

253

Cortisol 154
Creatinine excretion 20–1
Cyanogenic glycosides 102, 105, 109

Dactylopsila trivirgata 50, 60, 67
Dasycercus cristicauda 4, 7, 23, 42–4
Dasyuroides byrnei 4, 23, 38, 42–3, 108, 222
Dasyurus geoffroii 32, 34
Dasyurus hallucatus 4
Dasyurus maculatus 4, 34, 38–9, 54–5, 222
Dasyurus viverrinus 4, 15, 32, 34, 40, 108
Dendrolagus 115, 118, 166–7, 178
Dentition
  macropodines 159–60
  wombats 70, 71
Didelphis 3, 14, 28, 56–9, 67, 121–2
Diet
  arboreal folivores 65, 76, 81–3, 97, 101–6
  arboreal omnivores 27, 50, 59–64
  bandicoots 50–1
  caenolestids 29
  dasyurids 27–34, 64
  didelphids 56
  macropodines 159, 161–8, 192
  potoroines 200–4
  wombats 71
Digestion
  arboreal folivores 80, 86, 89–90, 101
  dasyurids 40–1
  macropodine stomach 143–4
Digestive tract
  arboreal folivores 76–8, 83–5, 97, 99, 102–3
  arboreal omnivores 65–6
  bandicoots 52–4
  caenolestids 37, 39
  carnivores 35
  dasyurids 38–9, 83–5
  didelphids 29, 36–7, 57
  macropodines 113, 115, 118
  potoroines 206
  wombats 71–3
Dorcopsis luctuosa 128–9, 135, 137, 167
Dorcopsulus 167
Dorre Island 51, 165, 167

Ecology
  Macropus eugenii 184–5, 188–91
  Macropus giganteus 192
  Macropus robustus erubescens 168–72
  Macropus rufus 172–6
  Pseudocheirus peregrinus 101
  rat-kangaroos 200
  Setonix brachyurus 176–84
  Thylogale thetis 195–7

Trichosurus caninus 81–3
Trichosurus vulpecula 76, 81–3
Endogenous urinary nitrogen 17–18
Energy, maintenance requirement
  birds 14
  eutherians 14
  marsupials 14–16, 41
Energy, sources to *Phascolarctos* 93–4
Essential oils 93–4, 106–9
Eucalyptus foliage as food resource 70, 75, 83, 88, 103–5, 212–13
Evolution
  foregut fermentation 112
  hindgut fermentation 69
  macropodid digestive tract 208–11
Feed intake 16–17, 86, 89, 97, 100
Fermentation, see Volatile fatty acids
Fibre digestion
  macropodines 142–4
  non-macropodid herbivores 80–1, 87–8
Fruit as food 61–2
Fungi
  as food 34, 200–4
  in macropodine stomach 141

Gas, in macropodine stomach 146–7
Gastric sulcus
  Macropodinae 113, 131–4
  Potoroinae 206–7
Glucokinase activity 153
Gluconeogenesis 109
Glucose 153–4
Glucuronic acid 94, 107, 221
L-Gulonolactone oxidase 221–5
Gymnobelideus leadbeateri 15–16, 50, 59–61, 65, 67

Haemoglobin 83, 171, 178
Heart rate, minimum 12
Heterothermic eutherians 8, 9
Hexokinase activity 153
Homeothermic eutherians 9
Homeothermy 1–3
Honeydew as food 65
Hypsiprymnodon moschatus 200, 204–6

Insulin 154
Intestinal disaccharidase activity 54–5, 152–3
Isoodon macrourus 5, 23, 54, 95, 222, 224
Isoodon obesulus 51, 54–5

Kangaroo Island 184, 188–91
Ketogenesis 152
Kidney function 58, 75, 81, 95–6, 156–8, 176, 185–6

Kidney, relative medullary thickness 59, 95, 178, 185–6, 198

Lagorchestes conspicillatus 165, 178
Lagorchestes hirsutus 165, 178
Lagostrophus fasciatus 167, 178
Lasiorhinus latifrons 6, 7, 19, 23–4, 70–5, 81
Lignin 90, 105–6
Lipid
 digestion in Koala 90, 93
 in *Eucalyptus* foliage 104–5
 metabolism in macropodines 155
Lysozyme activity in macropodine stomach 121

Macropus agilis 165
Macropus dorsalis 165
Macropus eugenii 5, 14–19, 21, 23, 86–7, 108, 116, 121, 124–35, 139–46, 149–58, 178–9, 184–91, 201, 216–17, 222
Macropus fuliginosus 70–1, 119, 192–6
Macropus giganteus 16, 19, 23, 55, 113–15, 119–39, 143–8, 151, 153, 155, 160–6, 178, 192–3, 207–8, 217–20, 222
Macropus irma 178
Macropus parma 131, 165, 198
Macropus parryi 165
Macropus robustus erubescens 14–19, 23, 95–6, 111, 146, 161–4, 168–76, 217–20
Macropus robustus robustus 19, 23, 131, 141, 146, 222
Macropus rufogriseus 16, 21, 32, 115, 117–18, 131, 141, 143, 146–8, 164, 222, 224
Macropus rufus 5, 14–18, 23, 75, 119–23, 128–32, 140, 142, 151, 154–5, 161–4, 172–8, 187, 210, 217–18
Macrotis lagotis 6, 7, 23, 51–3
Manna as food 64–5
Marmosa robinsoni 1, 29, 36, 37
Metabolic faecal nitrogen 87
Metacheirus nudicaudatus 3
Methane, in macropodine stomach 146–7
Microbial protein synthesis 150–1, 155
Microbiology, of macropodine stomach 140–1
Molar progression 165–6
Molybdenum 213–14
Monotremes 1, 2, 8, 37, 222–4
Myrmecobius fasciatus 1, 28, 30, 38

Nectar as food 62
Nitrogen balance 14, 47, 48, 173
Nitrogen, maintenance requirement
 eutherians 17, 19
 marsupials 19, 20, 193, 198
Notoryctes typhlops 28, 30, 38

Nutrition
 Didelphis 58
 Koala 86
 Quokka 179–84

Omnivore diets 50
Onychogalea fraenata 165
Ornithorhynchus anatinus (Platypus) 2, 8, 222–3

Peradorcus concinna 131, 165, 168, 178
Perameles bougainville 51
Perameles eremiana 51
Perameles gunnii 51
Perameles nasuta 5, 23, 52, 54–5, 67, 222, 224
Petauroides volans 75, 91, 97–9, 213, 222
Petaurus australis 50, 60–7
Petaurus breviceps 5, 16, 50, 60–7, 108–9
Petrogale inornata 23, 160, 168
Petrogale xanthopus 162–8
Phalanger maculatus 6, 7, 62, 75
Phalanger orientalis 62, 75
Phascogale tapoatafa 4, 38, 39, 64
Phascolarctos cinereus 7, 19, 55, 75, 83–96, 102, 104, 106–8, 213
Phenolics 90, 93, 94, 106–9
Philander opossum 4, 38, 39, 64
Phosphorus 194–5
Planigale spp. 28, 30
Planigale gilesi 44–5
Planigale ingrami 44–5
Planigale maculatus 4
Planigale tenuirostris 44
Pollen as food 62
Potassium 217–20
Potorous tridactylus 13, 20–1, 23, 108, 200–6
Protein turnover in liver
 eutherians 13
 Sminthopsis crassicaudata 13
Pseudocheirus peregrinus 55, 62, 75, 97, 101–3, 213, 222

Rate of passage
 arboreal folivores 78–80, 88–9, 101
 bandicoots 54, 56
 dasyurids 40
 macropodines 122–6, 134–5
 wombats 73–4
Regurgitation, in macropodines 117–19
Reproduction 46–9, 174–5, 177
Ribonuclease activity in pancreas 121
Rottnest Island 177–84, 215–17

Salivary glands
 didelphids 56

Salivary glands (*cont.*)
  macropodines 199–21, 220
  wombats 220
Sarcophilus harrisii 4, 15, 28, 32–3, 40
Scat analysis
  carnivores 28, 30
  didelphids 56
Sea water 185–6
Seeds as food 62
Selenium 215–16
Setonix brachyurus 16–18, 21, 113, 122, 128, 140–2, 146–7, 152–4, 161, 166–7, 176–84, 213–17
Sminthopsis crassicaudata 4–7, 13–14, 23–6, 30–1, 42–5, 49, 108
Sminthopsis macroura 43–4
Snowy Mountains 217–20
Sodium 217–20
Squamous epithelium, macropodine stomach 135–6
Stomach
  arboreal folivores 76
  caenolestids 35, 36
  didelphids 57
  macropodines 112–14, 116, 127–30
  potoroines 204–7
  wombats 71
Stomach, gastric histology of macropodines 135–40
Sulphur balance 173

Tachyglossus aculeatus (Echidna) 1, 8, 222–3
Tannins 88, 106–7
Tarsipes spencerae 27, 62–3, 66–7
Thylacinus cynocephalus 28, 30, 38
Thylogale billardierii 32, 117, 129, 137, 196–7, 208

Thylogale brunii 195, 197
Thylogale stigmatica 129, 137, 195, 197, 208
Thylogale thetis 16–21, 111, 124–9, 133–43, 147–51, 161, 167, 178, 195–8, 208, 216
Thyroxine 6, 12–14
Tolerance to 1080 in Brushtails 81
Torpor 25, 41–5
Transaminase activity in carnivores 28
Trichosurus caninus 75, 81–3
Trichosurus vulpecula 5, 19, 23, 55, 75–83, 87–8, 101, 105–8, 154, 222

Urea excretion 43, 58
Urea recycling 87, 155–6
Urea utilisation 156, 179–80, 188
Urine concentrating ability 43, 59, 75, 81, 96
UR ratio 86–7, 157, 181, 187–9, 207–8

Vitamin B_{12} 215
Vitamin E 216–17
Volatile fatty acids (VFA)
  concentration in macropodid stomach 141–3, 207
  metabolism in macropodines 151–4
  production in macropodines 147–9, 150, 193–5; in *Petauroides volans* 100; in *Phascolarctos* 90–3
Vombatus ursinus 70–5, 217–20, 222

Wallabia bicolor 131, 146–7, 160, 166, 222
Water metabolism in macropodines 175–6
Water turnover 21–2
  in eutherians 22–3
  in marsupials 22–6, 42–3, 75, 94–5
Wyulda squamicaudata 75